NON ADDITIVE GEOMETRY

NON ADDITIVE GEOMETRY

Shai Haran
Technion – Israel Institute of Technology, Israel

World Scientific

NEW JERSEY • LONDON • SINGAPORE • BEIJING • SHANGHAI • TAIPEI • CHENNAI

Published by

World Scientific Publishing Co. Pte. Ltd.
5 Toh Tuck Link, Singapore 596224
USA office: 27 Warren Street, Suite 401-402, Hackensack, NJ 07601
UK office: 57 Shelton Street, Covent Garden, London WC2H 9HE

Library of Congress Control Number: 2025005554

British Library Cataloguing-in-Publication Data
A catalogue record for this book is available from the British Library.

NON ADDITIVE GEOMETRY

ISBN 978-981-98-0668-3 (hardcover)
ISBN 978-981-98-0669-0 (ebook for institutions)
ISBN 978-981-98-0670-6 (ebook for individuals)

For any available supplementary material, please visit
https://www.worldscientific.com/worldscibooks/10.1142/14145#t=suppl

Desk Editors: Soundararajan Raghuraman/Rok Ting Tan

Typeset by Stallion Press
Email: enquiries@stallionpress.com

This monograph is dedicated to the memory of my teachers, mentors, and friends

Daniel Quillen
Yuri Manin
John Coates

Preface

The usual dictionary between geometry and commutative algebra is not appropriate for Arithmetic geometry because addition is a singular operation at the "Real prime". We replace Rings, with addition and multiplication, by Props (=strict symmetric monoidal category generated by one object), or by Bioperad (=two closed symmetric operads acting on each other, a mathematical formulation of Dirac's Bra-Ket notation): to a ring we associate the prop of all matrices over it, with matrix multiplication and block direct sums as the basic operations, or the bioperad consisting of all row and column vectors over it. We define the "commutative" props and bioperads, and using them we develop a generalized algebraic geometry, following Grothendieck footsteps closely. This new geometry is appropriate for Arithmetic (and potentially also for Physics).

About the Author

Shai Haran graduated from the Hebrew University in Jerusalem and received his PhD in mathematics from the Massachusetts Institute of Technology (MIT) on "p-Adic L-functions for Modular forms" under his advisor Barry Mazur from Harvard University and his mentors Michael Artin and Daniel Quillen from MIT.

Haran is a professor at the Technion — Israel Institute of Technology. He was a frequent visitor at Stanford University, the various universities of California, MIT, Harvard and Columbia University, USA; Cambridge and Oxford universities, England; the Institut des Hautes Études Scientifiques, Paris, France; the Max-Planck Institute, Bonn, Germany; Kyushu University, and the Tokyo Institute of Technology, Japan, among other institutions.

Haran is a pacifist and a conscious objector.

Contents

Introduction

We, the human beings, are walking on this earth doing most of the time addition and subtraction in our brains — "How much will I profit? earn? pay? How much it will cost? What will be left?". There's no wonder that addition and subtraction dominate all our mathematics. It is always an Abelian group that is at the basis of any mathematical structures (or an additive functor between Abelian categories, and homological algebra that produces all the deeper theorems). However, there is a world before addition, before we apply the functor of "the free Abelian group" and start doing addition and subtraction — it is the world of homotopy. It is the long exact sequences of (co)fibrations that give us all the deeper theorems we need.

The encoding of geometry into an Abelian group begins with Gelfand theorem on the equivalence of the category of compact Hausdorff topological spaces and continuous maps, with the opposite of the category of commutative C^*-algebra and $*$-homomorphism. To a topological space X, one associate the commutative C^*-algebra $C(X)$, defined as

$$C(X) = \{f : X \to \mathbb{C} \text{ continuous}\}$$

and use the addition, multiplication (conjugation, and norm) in $\mathbb{C}$ to form the structure of a C^*-algebra on $C(X)$. To a commutative C^*-algebra A one associate the space $\mathit{spec}(A)$ of maximal ideals of A, which is compact Hausdorff with respect to the minimal topology making all $f \in A$ continuous. The axiom of C^*-algebras are a direct generalization of the axioms of the complete field of complex numbers; indeed for a point $X = \{*\}$, $C(*) = \mathbb{C}$.

Alexander Grothendieck (Shapiro) came to algebraic geometry from C^*-algebras, and in front of him were the Weil conjectures which are all "above $\mathbb{Z}$". He therefore chose **Commutative Rings** as the basis for the language of algebraic geometry. If he was more interested in Arithmetic he would have had to choose a more general notion than that of a commutative ring. That commutative rings are not sufficient for Arithmetic geometry is clear by considering Weil's "Rosetta stone", that's the one dimensional structures we have in the mathematical reality:

Number Fields: $K/\mathbb{Q}$ finite. (1*)

Function Fields in characteristic p: $K/\mathbb{F}_q(z)$ finite. (2*)

Function Fields in characteristic 0: $K/\mathbb{C}(z)$ finite. (3*)

The Function Fields K corresponds one to one with smooth projective curves X_K, and the points of X_K are in bijection with the valuation sub rings of K. A choice of the transcendental basis "z" corresponds to a choice of a dominant mapping $z : X_K \twoheadrightarrow \mathbb{P}^1$.

At the basis of the Rosetta stone we have the rational numbers/functions, with their real prime/point at infinity: see Figure 0.1.

We see that the "Real integers" $\mathbb{Z}_{\mathbb{R}} = [-1, 1] \subseteq \mathbb{R}$, and similarly the "Complex integers" $\mathbb{Z}_{\mathbb{C}} = \{z \in \mathbb{C},\ |z| \leqslant 1\}$, are not closed under addition, and do not form a sub ring. As Grothendieck knew well: the global theorems exist only in projective geometry, not in affine geometry. The most basic example being that a rational function without poles anywhere, including the point at infinity, must be a constant. If our language does not "see" all of the points there will be no global theorems. We must therefore understand these real and complex "integers" as some kind of a "generalized ring".

Another problem is the mysterious "Arithmetic surface": the integers $\mathbb{Z}$ are the initial object of the category of rings, so the categorical sum of $\mathbb{Z}$ with itself reduces to the diagonal

$$\mathbb{Z} \otimes \mathbb{Z} = \mathbb{Z}, \tag{0.1}$$

unlike the function field analog

$$F[z] \otimes_F F[z] = F[z_1, z_2]. \tag{0.2}$$

$$\mathbb{Z}_{\mathbb{C}} \equiv \{z \mid |z| \leqslant 1\} \subseteq \mathbb{C}$$
$$\cup| \qquad\qquad \cup|$$
$$\mathbb{Z}_{\mathbb{R}} \equiv [-1,1] \subseteq \mathbb{R}$$

(1*):
$$\mathbb{Z} \subseteq \mathbb{Q}$$
$$\mathbb{F}_p \twoheadleftarrow \mathbb{Z}_p \subseteq \mathbb{Q}_p$$

(2*):
$$\mathbb{F}_q \twoheadleftarrow \mathbb{F}_q\left[\left[\tfrac{1}{z}\right]\right] \subseteq \mathbb{F}_q\left(\left(\tfrac{1}{z}\right)\right)$$
$$\mathbb{F}_q[z] \subseteq \mathbb{F}_q(z)$$
$$\mathbb{F}_{q^{\deg f}} \twoheadleftarrow \mathbb{F}_{q^{\deg f}}[[f]] \subseteq \mathbb{F}_{q^{\deg f}}((f))$$

(3*):
$$\mathbb{C} \twoheadleftarrow \mathbb{C}\left[\left[\tfrac{1}{z}\right]\right] \subseteq \mathbb{C}\left(\left(\tfrac{1}{z}\right)\right)$$
$$\mathbb{C}[z] \subseteq \mathbb{C}(z)$$
$$\mathbb{C} \twoheadleftarrow \mathbb{C}[[z-\alpha]] \subseteq \mathbb{C}((z-\alpha))$$

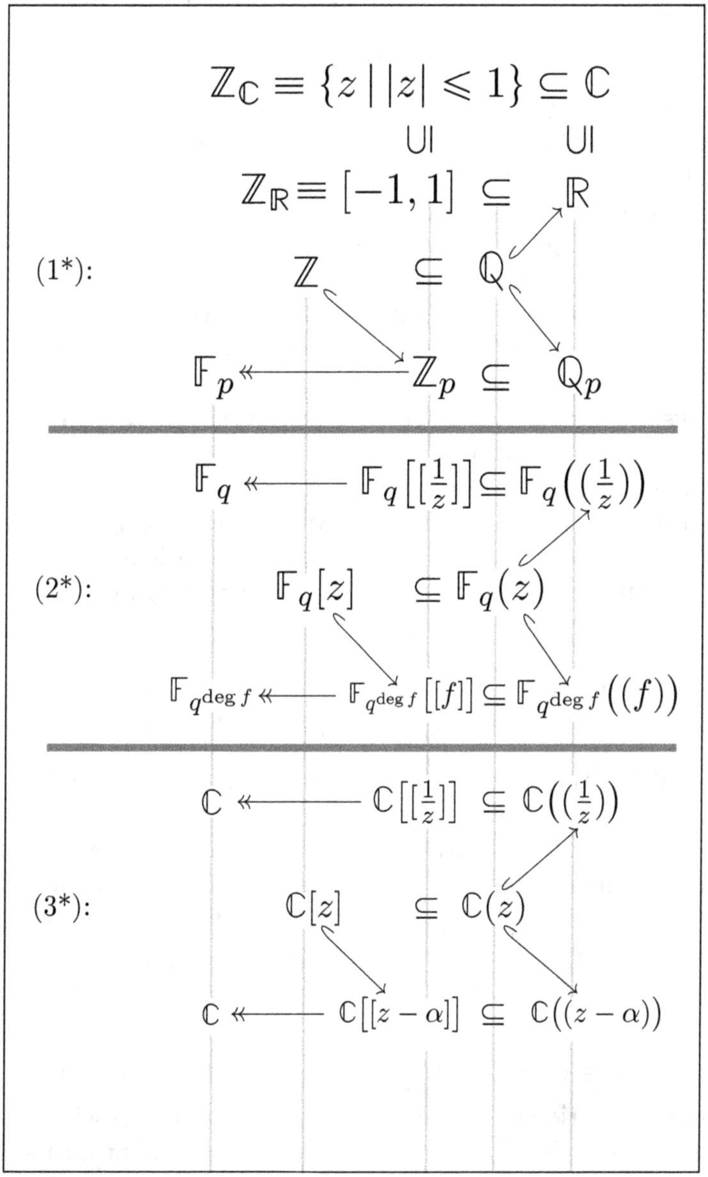

Figure 0.1. The mathematical Rosetta stone.

Thus if we insist on rings, the **Arithmetical Surface**

$$\mathcal{spec}(\mathbb{Z}) \prod \mathcal{spec}(\mathbb{Z}) = \mathcal{spec}(\mathbb{Z}) \tag{0.3}$$

reduces to its diagonal.

The original Rosetta Stone had three languages talking word for word about one and the same reality. Comparing the well understood ancient Greek and Demotic languages with the mysterious Hieroglyphic, the ancient Egyptian language, led to the decipherment of the Hieroglyphics, and with it the understanding of many texts from the mummies and the pyramids. While in mathematics we have one language that is talking about three "one dimensional" but different realities (1*), (2*), and (3*). Moreover, this language of commutative rings, the language of **Addition and Multiplication**, is not the right language for Arithmetic geometry, because addition is singular at the real prime, and the Arithmetic Surface disappears.

A similar situation occurs in physics: normalizing the speed of light to one, $c = 1$, the interval of speeds $[-1, 1]$ is not closed under addition. Einstein's solution was to replace $z_1 + z_2$ by

$$z_1(+)z_2 := (z_1 + z_2)/(1 + z_1 \bullet z_2) \tag{0.4}$$

That is we use the fractional linear transformation

$$z \mapsto (1 + z)/(1 - z)$$

that identifies the interval $[-1, 1]$ with the positive reals $[0, \infty]$, and we carry multiplication of positive real numbers $(0, \infty)$ to the interval $(-1, 1)$ to obtain Einstein's addition $(+)$. When we do it with complex numbers, the unit disc $\mathbb{Z}_{\mathbb{C}} = \{|z| \leqslant 1\}$ get identified with the right half plane $\{Re(z) \geqslant 0\}$ which is not closed with respect to multiplication, so $\mathbb{Z}_{\mathbb{C}}$ is not closed under $(+)$.

(Curiously, we do have that $\mathbb{Z}_{\mathbb{C}}$ is closed with respect to the operations

$$z_1(+)\text{'}z_2 := (z_1 + \overline{z}_2)/(1 + z_1 z_2), \quad \text{or} \quad (z_1 + z_2)/(1 + z_1\overline{z}_2)$$

but these operations are neither commutative nor associative).

It is perhaps already here that relativity and quantum mechanics cannot coexist. Quantum mechanics started with the observation by Heisenberg that the level structure of the energy of microscopic particles are controlled by finite matrices.

Our basic idea is to replace the ring A by the collection of all matrices over $A = \{A_{n,m}\}$, $n, m \geqslant 0$ with the operations of

Matrix Multiplication and of Block Direct Sum of Matrices

The algebraic structure of this collection of all matrices is that of a "**Prop**":

a strict symmetric monoidal category generated by one object

Observing that

$$(\mathbb{Z}_p)_{n,m} = \{a \in (\mathbb{Q}_p)_{n,m} \mid a \circ \mathbb{Z}_p^m \subseteq \mathbb{Z}_p^n\} \tag{0.5}$$

we define

$$(\mathbb{Z}_\mathbb{R})_{n,m} = \{a \in (\mathbb{R})_{n,m} \mid a \circ \mathbb{Z}_\mathbb{R}^m \subseteq \mathbb{Z}_\mathbb{R}^n\} \tag{0.6}$$

with

$$\mathbb{Z}_\mathbb{R}^m = \{(z_1, \ldots, z_m) \in \mathbb{R}^m, \quad |z_1|^2 + \cdots + |z_m|^2 \leqslant 1\} \tag{0.7}$$

the unit ℓ_2 ball. Similarly the complex "integers" are defined by

$$(\mathbb{Z}_\mathbb{C})_{n,m} = \{a \in \mathbb{C}_{n,m} \mid a \circ \mathbb{Z}_\mathbb{C}^m \subseteq \mathbb{Z}_\mathbb{C}^n\} \tag{0.8}$$

with

$$\mathbb{Z}_\mathbb{C}^m = \{(z_1, \ldots, z_m) \in \mathbb{C}^m, |z_1|^2 + \cdots + |z_m|^2 \leqslant 1\} \tag{0.9}$$

the unit ℓ_2 complex ball. The sub props

$$Z_\mathbb{R} \subseteq \mathbb{R}, \qquad Z_\mathbb{C} \subseteq \mathbb{C} \tag{0.10}$$

are closed under the operations of matrix multiplication and block direct sum of matrices (and just like $\mathbb{Z}_p \subseteq \mathbb{Q}_p$, are maximal compact "valuation" sub props).

We can define the "residue field at the real prime" $\mathbb{F}_\mathbb{R}$, and similarly the "residue field at the complex prime" $\mathbb{F}_\mathbb{C}$, as follows. For $a \in (\mathbb{Z}_\mathbb{R})_{n,m}$ we have $a^t \in (\mathbb{Z}_\mathbb{R})_{m,n}$, and the self-adjoint operators $a^t \circ a \in (\mathbb{Z}_\mathbb{R})_{m,m}$, $a \circ a^t \in (\mathbb{Z}_\mathbb{R})_{n,n}$. The spectral decomposition of these operators give

$$\mathbb{R}^n = \ker(a^t) \oplus \bigoplus_{i=1}^{k} W(\lambda_i) \qquad \mathbb{R}^m = \ker(a) \oplus \bigoplus_{i=1}^{k} V(\lambda_i) \tag{0.11}$$

with $0 < \lambda_1 < \lambda_2 < \cdots < \lambda_k \leqslant 1$, and a induces isomorphism

$$W(\lambda_i) = \ker\left(a \circ a^t - \lambda_i \circ \mathrm{Id}_n\right) \xleftarrow[\sim]{a} V(\lambda_i) = \ker\left(a^t \circ a - \lambda_i \cdot \mathrm{Id}_m\right) \tag{0.12}$$

In particular, a induces the partial isometry (possibly empty!)

$$\widehat{a} := \{W(1) \xleftarrow[\sim]{a} V(1)\} \tag{0.13}$$

Let $\mathbb{F}_{\mathbb{R}}$ denote the collection of all such partial isometries

$$(F_{\mathbb{R}})_{n,m}$$
$$:= \{\mathcal{W}(a) \xleftarrow[\sim]{a} \mathcal{V}(a) \text{ linear isometry}, \mathcal{W}(a) \subseteq \mathbb{R}^n, \mathcal{V}(a) \subseteq \mathbb{R}^m\}$$

The composition in $\mathbb{F}_{\mathbb{R}}$ is given by

$$(\mathcal{W}(a) \xleftarrow[\sim]{a} \mathcal{V}(a)) \circ (\mathcal{W}(b) \xleftarrow[\sim]{b} \mathcal{V}(b))$$
$$:= (a(\mathcal{V}(a) \cap \mathcal{W}(b)) \xleftarrow[\sim]{a \circ b} b^{-1}(\mathcal{V}(a) \cap \mathcal{W}(b))) \tag{0.14}$$

The direct sum in $\mathbb{F}_{\mathbb{R}}$ is the usual sum

$$((\mathcal{W}(a_1)) \xleftarrow[\sim]{a_1} \mathcal{V}(a_1)) \oplus (\mathcal{W}(a_2) \xleftarrow[\sim]{a_2} \mathcal{V}(a_2))$$
$$:= (\mathcal{W}(a_1) \oplus \mathcal{W}(a_2) \xleftarrow[\sim]{a_1 \oplus a_2} \mathcal{V}(a_1) \oplus \mathcal{V}(a_2)) \tag{0.15}$$

With these operations $\mathbb{F}_{\mathbb{R}}$ forms a prop, and we have the surjection

$$\begin{aligned} \mathbb{Z}_{\mathbb{R}} &\twoheadrightarrow \mathbb{F}_{\mathbb{R}} \\ a &\longmapsto \hat{a} := (W(1) \xleftarrow[\sim]{a} V(1)) \end{aligned} \tag{0.16}$$

In exactly the same way (using $\overline{a}^t$ instead of a^t) we have the surjection of props

$$\mathbb{Z}_{\mathbb{C}} \twoheadrightarrow \mathbb{F}_{\mathbb{C}} \tag{0.17}$$

With these definitions, we see that the "mathematical Rosetta stone" can be repaired, and in the language of props, the Arithmetic reality

(1*) becomes word for word analogue to the function fields realities (2*) and (3*):

$$
\begin{array}{ccccccc}
 & & & \mathbb{C} & \supseteq & \mathbb{Z}_{\mathbb{C}} & \twoheadrightarrow \mathbb{F}_{\mathbb{C}} \\
 & & & \rotatebox[origin=c]{90}{$\subseteq$} & & \rotatebox[origin=c]{90}{$\subseteq$} & \quad \rotatebox[origin=c]{90}{$\subseteq$} \\
 & & & \mathbb{R} & \supseteq & \mathbb{Z}_{\mathbb{R}} & \twoheadrightarrow \mathbb{F}_{\mathbb{R}} \\
 & & & \rotatebox[origin=c]{90}{$\subseteq$} & & & \\
 & \mathbb{Z} & \subseteq & \mathbb{Q} & & & \\
 & \rotatebox[origin=c]{-90}{$\subseteq$} & & \rotatebox[origin=c]{-90}{$\subseteq$} & & & \\
\mathbb{F}_p \twoheadleftarrow & \mathbb{Z}_p & \subseteq & \mathbb{Q}_p & & &
\end{array}
\tag{0.18}
$$

Notice that the prop associated to any ring A will contain the sub prop $\mathbb{F}$ with

$$
\mathbb{F}_{n,m} = \boxed{\begin{array}{l}\text{all the n by m matrices with entries 0 or 1 such that}\\ \text{in any row and in any column there will be at most}\\ \text{one nonzero entry.}\end{array}}
\tag{0.19}
$$

This common sub prop of all rings is the "**Field with One element**" [**Sou08**]. It is the initial object of the category of props, and the "**absolute-point**" $\mathcal{spec}(\mathbb{F})$ will be the last object of our geometry.

There is no such thing as non commutative geometry !

Although non commutative rings (and similarly non-commutative props) have a rich theory of cohomology and K-theory generalizing the theory of commutative rings [**Con94**], there is no spectrum associated with a non commutative algebra, there are no spaces, localization and sheaves. We will have to understand what it means for the prop A when the underlying ring A is commutative. There will be (at least) two notions of commutativity for props giving rise to the full embeddings beginning with the category of commutative rings $\mathcal{CRing}$,

$$
\mathcal{CRing} \subseteq \mathcal{C}_T\mathcal{Prop} \subseteq \mathcal{CProp} \subseteq \mathcal{Prop}
\tag{0.20}
$$

We refer to the objects of $\mathcal{CProp}$ as the "commutative props", and to those of $\mathcal{C}_T\mathcal{Prop}$ as the "totally commutative props".

These categories are all complete and co-complete.

In particular we have the "Arithmetical Surface", the categorical sum of $\mathbb{Z}$ with itself in the category $\mathcal{CProp}$, which we denote by $\mathbb{Z} \otimes_{\mathbb{F}} \mathbb{Z}$. Since $\mathbb{Z}$ as a prop is totally-commutative, the diagonal map factors through the totally-commutative quotient

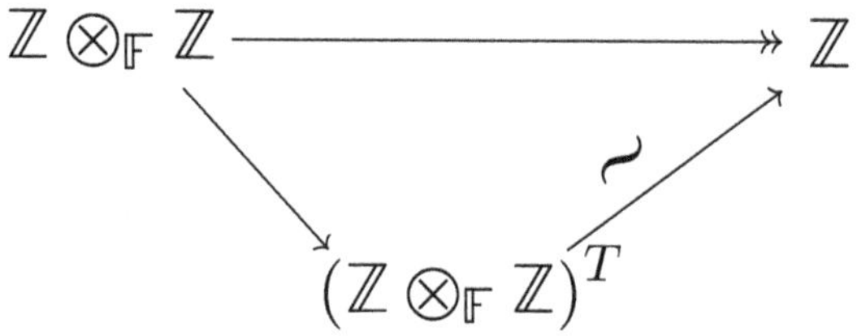

(0.21)

Unfortunately, the categorical sum in the category $\mathcal{C}_T\mathcal{Prop}$ of $\mathbb{Z}$ with itself, the total commutative quotient of $\mathbb{Z} \otimes_{\mathbb{F}} \mathbb{Z}$, reduces to $\mathbb{Z}$. This is why we prefer to work with the weaker notion of commutativity.

To a commutative prop A we will associate functorially a compact topological space $\mathcal{spec}(A)$. This space depends on much less information than what A gives. To construct $\mathcal{spec}(A)$ we only need the information given by the collection of "row and column vectors in A", we call this UA, so

$$(\mathrm{U}A)^-(n) := A_{1,n}, \quad (\mathrm{U}A)^+(n) = A_{n,1} \tag{0.22}$$

The collection $\mathrm{U}A = \{\mathcal{P}^-(n), \mathcal{P}^+(n)\}$ forms the structure we call **Bio**, short for **Bi-operad**, consisting of two (closed symmetric) operads $\mathcal{P}^-$ and $\mathcal{P}^+$ acting on each other in a consistent manner.

We obtain a commutative diagram with rows full embeddings and their left adjoints given by taking the (respectively, totally) commutative quotient

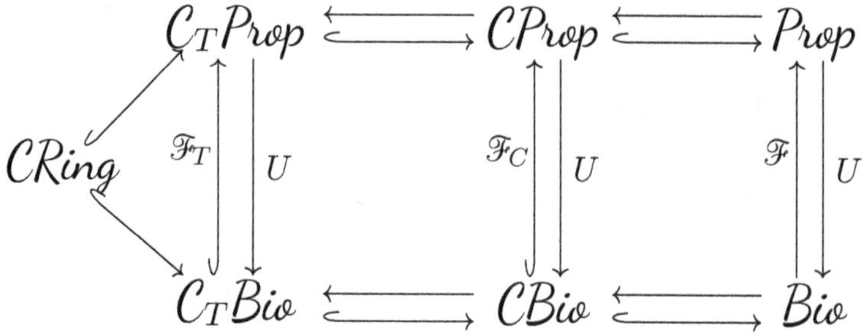

(0.23)

The functor $\mathcal{F}$ (respectively, $\mathcal{F}_T$, $\mathcal{F}_C$) is the Free (respectively, totally, commutative) prop generated by the given bio.

Here $\mathcal{CBio}$ are the bios $\mathcal{P} = (\mathcal{P}^-, \mathcal{P}^+)$ where all operations $p \in \mathcal{P}^-(n)$, and all co-operations $q \in \mathcal{P}^+(m)$, commute in the sense of bi-algebras:

$$q \circ p = (\underbrace{p, \ldots, p}_{m}) \circ \sigma_{m,n} \circ (\underbrace{q, \ldots, q}_{n}) \tag{0.24}$$

Thus we identify

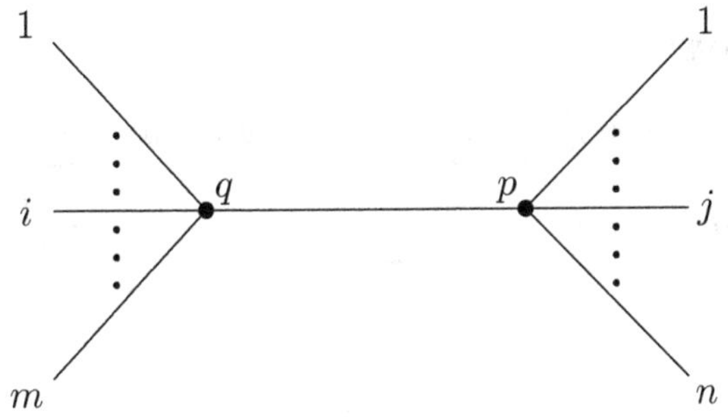

(0.25)

with

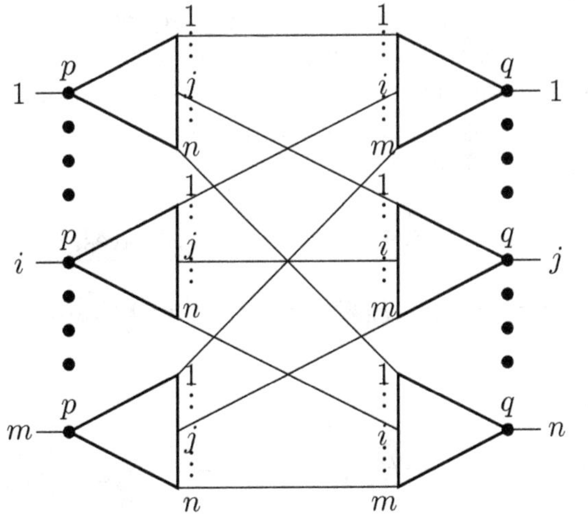

(0.26)

$$(i-1)\cdot n + j = \sigma_{m,n}((j-1)m + i), \quad 1 \leqslant i \leqslant m, \quad 1 \leqslant j \leqslant n \tag{0.27}$$

the permutation $\sigma_{m,n} \in S_{m \bullet n}$ is given pictorially in the diagram (0.26).

Having an n by m matrix we can order its entries row by row, or column by column, and these two ways of ordering the set with $n \cdot m$ elements give us the permutation $\sigma_{m,n} \in S_{m \cdot n}$. These permutations will be used throughout this book. Note that

$$\sigma_{n,m} = (\sigma_{m,n})^{-1}, \sigma_{n,1} = \sigma_{1,n} = \mathrm{Id}_n$$

The totally commutative bios $\mathcal{C}_T\mathcal{B}io$ are the commutative bios $P = (P^-, P^+)$ where the operads P^- and P^+ satisfy further the original Boardman–Vogt commutativity:

$$\begin{aligned}
&\text{for } p \in P^-(m),\ p' \in P^-(n) : p \circ (\underbrace{p', \ldots, p'}_{m}) = p' \circ (\underbrace{p, \ldots, p}_{n}) \circ \sigma_{n,m} \\
&\text{for } q \in P^+(m),\ q' \in P^+(n) : (\underbrace{q', \ldots, q'}_{m}) \circ q = \sigma_{m,n} \circ (\underbrace{q, \ldots, q}_{n}) \circ q'
\end{aligned} \tag{0.28}$$

For a commutative bio $P = (P^-, P^+)$, (respectively, prop A), the monoid $P^-(1) \cong P^+(1)$ $(= A_{1,1})$ is commutative and central, so that we can localize with respect to multiplicative subsets $S \subseteq P(1)(= A_{1,1})$. We obtain a sheaf $\mathcal{O}_P$ (respectively, $\mathcal{O}_A$) of commutative bios (respectively, props) with local stalks over $\mathcal{S}pec(P)$ (respectively, $\mathcal{S}pec(\mathrm{U}A)$). To a map of bios (props) φ corresponds a map of locally bio (respectively, prop) spaces in the opposite direction. We obtain the diagram of full-embedding

$$\begin{array}{ccccc}
\underline{\text{Affine}} & & \underline{\text{Schemes}} & & \underline{\text{Sheaves, local stalks}} \\
(\mathcal{CR}ing)^{\mathrm{op}} & \hookrightarrow & \mathcal{S}ch & \hookrightarrow & \mathcal{CR}ing/\mathcal{T}op \\
\downarrow & & \downarrow & & \downarrow \\
(\mathcal{CB}io)^{\mathrm{op}} & \hookrightarrow & \mathcal{BS}ch & \hookrightarrow & \mathcal{CB}io/\mathcal{T}op \\
\downarrow & & \downarrow & & \downarrow \\
(\mathcal{CP}rop)^{\mathrm{op}} & \hookrightarrow & \mathcal{PS}ch & \hookrightarrow & \mathcal{CP}rop/\mathcal{T}op
\end{array} \tag{0.29}$$

Here the column on the right of "**sheaves with local stalks**", consists of pairs $(X, \mathcal{O}_X)$ of a topological space X together with a sheaf $\mathcal{O}_X$ of $\mathcal{CRing}/\mathcal{CBio}/\mathcal{CProp}$ with local stalks at each point $x \in X$

$$\mathcal{O}_{X,x} := \varinjlim_{x \in \mathcal{U} \subseteq X} \mathcal{O}_X(\mathcal{U}) \tag{0.30}$$

with maps $f : X \to Y$ are pairs of a continuous map f and pull-back of sections $f_{\mathcal{U}}^{\#} : \mathcal{O}_Y(\mathcal{U}) \to \mathcal{O}_X(f^{-1}\mathcal{U})$ compatible with restriction with respect to the open subsets $\mathcal{U} \subseteq Y$, and inducing **local** maps on stalk $f_x^{\#} : \mathcal{O}_{Y,f(x)} \to \mathcal{O}_{X,x}$, $f_x^{\#}(\mathfrak{m}_{Y,f(x)}) \subseteq \mathfrak{m}_{X,x}$.

The middle column of "**schemes**", consists of the category of ordinary schemes $\mathcal{Sch}$, the category of $\mathcal{Bio}$ schemes $\mathcal{BSch}$, and the category of $\mathcal{Prop}$ schemes $\mathcal{PSch}$. These are the full-subcategories of the right column consisting of $(X, \mathcal{O}_X)$ which are locally affine:

we have $X = \bigcup_\alpha \mathcal{U}_\alpha$, $\mathcal{U}_\alpha$ open, and

$$(\mathcal{U}_\alpha, \mathcal{O}_X|_{\mathcal{U}_\alpha}) = \mathit{spec}(\mathcal{O}_X(\mathcal{U}_\alpha)) \quad \text{all } \alpha \tag{0.31}$$

These categories of generalized scheme are finite-complete and we can always construct fiber products

$$\begin{array}{ccccc} & & X_0 \prod\limits_Y X_1 & & \\ & \swarrow & & \searrow & \\ X_0 & & & & X_1 \\ & {\scriptstyle f_0}\searrow & & \swarrow{\scriptstyle f_1} & \\ & & Y & & \end{array} \tag{0.32}$$

But we will be interested in more general limits of the form $\varprojlim_{j \in J} X_j$, where J is a partially ordered set, and for simplicity we restrict to

$$J \ \underline{\text{directed}}: \forall j_1, j_2 \in J,\ \exists j \in J,\ j \geqslant j_1 \ \& \ j \geqslant j_2 \tag{0.33}$$

$$\underline{\textit{co–finite}}: \forall j \in J,\ \#\{j' \in J, j' \leqslant j\} < \infty \tag{0.34}$$

and where $j \rightsquigarrow X_j$ is a functor, so that for $j \geqslant j'$ we have a map $X_j \to X_{j'}$.

These limits always exist in the categories of the right column: the underlying space of $\lim_{\overleftarrow{j\in J}} X_j$ is the set of all coherent sequences of points $x = \{x_j\}_{j\in J}$, with the inverse limit topology, and the sheaf is the direct limit of $\mathcal{O}_{X_j}$ (and $\mathcal{O}_{\lim_{\leftarrow} X_j}|_{\{x_j\}} = \lim_{\rightarrow} \mathcal{O}_{X_j,x_j}$ is local as the colimit of local homomorphisms).

Similarly, such limits always exists in the categories of the left column of Affine Schemes. Indeed we have quite formally

$$\lim_{\overleftarrow{j\in J}} \mathit{spec}(A_j) = \mathit{spec}\left(\lim_{\overrightarrow{j\in J}} A_j\right) \tag{0.35}$$

But the categories of schemes, and of generalized schemes, do not have such limits: for a coherent sequence of points

$$x = \{x_j\}_{j\in J} \in \lim_{\overleftarrow{j\in J}} X_j$$

while each $x_j \in X_j$ has an affine neighborhood $\mathcal{U}_j \subseteq X_j$, the intersection $\bigcap_{j\in J} \pi_j^{-1}(\mathcal{U}_j)$ need not be open.

We therefore pass to the pro-categories of (generalized) schemes, with objects arbitrary such inverse systems $\{X_j\}_{j\in J}$, and maps

$$\mathit{pro}\text{-}\mathcal{S}ch\left(\{X_j\}_{j\in J}, \{Y_i\}_{i\in I}\right) := \lim_{\overleftarrow{i\in I}} \lim_{\overrightarrow{j\in J}} \mathcal{S}ch\,(X_j, Y_i) \tag{0.36}$$

In the categories $\mathit{pro}\text{-}\mathcal{PS}ch$ and $\mathit{pro}\text{-}\mathcal{BS}ch$ we finally have the compactification of $\mathit{spec}(\mathbb{Z})$, denoted $\overline{\mathit{spec}(\mathbb{Z})}$, and similarly for any number field K we have $\overline{\mathit{spec}(\mathcal{O}_K)}$. For $\overline{\mathit{spec}(\mathbb{Z})}$ we take J to be the collection of finite subsets of the primes, with inclusion as partial order, and

$$X_{\{p_1\cdots p_\ell\}} := \mathit{spec}(\mathbb{Z}) \coprod_{\mathit{spec}\left(\mathbb{Z}\left[\frac{1}{p_1\cdots p_\ell}\right]\right)} \mathit{spec}\left(\mathbb{Z}\left[\frac{1}{p_1\cdots p_\ell}\right] \cap \mathbb{Z}_{\mathbb{R}}\right) \tag{0.37}$$

The generalized scheme $X_{\{p_1,\ldots,p_\ell\}}$ has closed points

$$\{(p_1), \ldots, (p_\ell), \eta_{\mathbb{R}}\}$$

with $\eta_{\mathbb{R}} = \mathbb{Z}[\frac{1}{p_1\cdots p_\ell}] \cap (-1,1)$, the "**Real prime**", coming from the unique maximal ideal $(-1,1)$ of $\mathbb{Z}_{\mathbb{R}}$. Enlarging the set of primes we have maps

$$X_{\{p_1,\ldots,p_\ell,q_1,\ldots,q_k\}} \longrightarrow X_{\{p_1,\ldots p_\ell\}} \tag{0.38}$$

these maps are identity on points, and are identity on the sheaves, but there are more open neighborhoods of the real prime $\eta_{\mathbb{R}}$ in $X_{\{p_1,\ldots p_\ell,q_1,\ldots,q_k\}}$ then there are in $X_{\{p_1,\ldots,p_\ell\}}$. The space

$$\varprojlim \overline{\mathit{spec}(\mathbb{Z})} \equiv \varprojlim_{\{\ell_1,\ldots \ell_\ell\}\in J} X_{\{\ell_1,\ldots,\ell_\ell\}} \tag{0.39}$$

is one-dimensional and has the generic point (0), and the closed points $\eta_{\mathbb{R}}$ and (p), where p is an arbitrary prime. The structure sheaf (as a prop) is given by

$$\mathcal{O}_{\varprojlim \overline{\mathit{spec}(\mathbb{Z})}}(\mathcal{U})_{n,m} = \left\{a \in \mathbb{Q}_{n,m}, a \circ \mathbb{Z}_p^m \subseteq \mathbb{Z}_p^n \text{ for all } p \in \mathcal{U}\right\} \tag{0.40}$$

with global sections

$$\mathcal{O}_{\varprojlim \overline{\mathit{spec}(\mathbb{Z})}}\left(\overline{\mathit{spec}(\mathbb{Z})}\right)_{n,m} = \mathbb{Z}_{n,m} \cap (\mathbb{Z}_{\mathbb{R}})_{n,m} = \mathbb{F}[\pm 1]_{n,m} \tag{0.41}$$

the n by m matrices where each row and each column will have 0 entries, accept at most in one entry which could be ± 1.

It is here that we obtain a truly global result: the eigenvalues of matrices in $\mathbb{F}[\pm 1]_{n,n}$ are either zero or roots of unity, the "constants" of arithmetic, and conversely: the algebraic integers having "no poles over the real prime" (in any embedding into $\mathbb{C}$ they are mapped into $\mathbb{Z}_{\mathbb{C}}$) are the roots of unity (Kronecker theorem).

For any prop A, and any $n \geqslant 1$, we have the monoid $A_{n,n}$, and its subgroup of invertible elements $\mathrm{GL}_n(A)$, the "general linear group" in dimension n.

For example,

$$\begin{aligned} \mathrm{GL}_n(\mathbb{F}) = S_n \subseteq \mathrm{GL}_n(\mathbb{F}[\pm 1]) &= (\pm 1)^n \rtimes S_n \\ &\subseteq \mathrm{GL}_n(\mathbb{Z}_{\mathbb{R}}) = O(n) \subseteq \mathrm{GL}_n(\mathbb{Z}_{\mathbb{C}}) = U(n) \end{aligned} \tag{0.42}$$

and $\mathrm{GL}_n(A)$ is the usual group of n by n invertible matrices for a ring A.

These groups come with associative maps

$$\oplus : \mathrm{GL}_{n_1}(A) \times \mathrm{GL}_{n_2}(A) \longrightarrow \mathrm{GL}_{n_1+n_2}(A) \tag{0.43}$$

and give the K-theoretic spectrum of A.

The groups $\mathrm{GL}_n(A)$ act via conjugation on $A_{n,n}$, and we have the associate orbit space

$$[A_{n,n}] := A_{n,n}/\mathrm{GL}_n(A)$$

On the direct limit as $n \to \infty$ we get a commutative monoid

$$[A] = \varinjlim_{n\to\infty} [A_{n,n}], \quad [a_1] + [a_2] := [a_1 \oplus a_2] \tag{0.44}$$

with an action of the multiplicative monoid of non-zero natural numbers by "Frobenius" operators

$$F_m[a] := [\underbrace{a \circ a \circ \cdots \circ a}_{m}], \quad F_{m_1} \circ F_{m_2} = F_{m_1 \cdot m_2} \tag{0.45}$$

Applying Grothendieck's K-functor localizing addition, we get an Abelian group

$$\mathcal{W}(A) := K[A] \tag{0.46}$$

When A is totally commutative, $A \in \mathcal{C}_T\mathcal{P}rop$, $\mathcal{W}(A)$ has multiplication via tensor product making it a commutative ring, and has λ-operations making it a special λ-ring.

"Applications are reserved to the future" — but we do give one new application taking $A = \mathcal{O}_{\overline{spec\mathbb{Z}}}(\overline{spec\mathbb{Z}}) = \mathbb{F}[\pm 1]$, we consider the λ-ring $\mathcal{W} = \mathcal{W}\left(\overline{spec(\mathbb{Z})}\right)$ which has an additive basis the cyclotomic polynomials ϕ_n, with ϕ_1 the unit of $\mathcal{W}$. We have ring homomorphisms by taking traces of matrices raised to the mth power

$$t_m = \mathrm{tr} \circ F_m : \mathcal{W} = \bigoplus_{n \geqslant 1} \mathbb{Z} \cdot \phi_n \twoheadrightarrow \mathbb{Z} \tag{0.47}$$

with

$$t_m(\phi_n) = \mathrm{tr}(\mathbb{F}_m \phi_n) = \sum_{\xi \in \mu_n^*} \xi^m = C_n^m = \mu\left(\frac{n}{(n,m)}\right) \frac{\varphi(n)}{\varphi\left(\frac{n}{(n,m)}\right)} \tag{0.48}$$

the **Ramanujan sums**.

This put the Ramanujan sums in a precise algebraic context.

The λ-ring $\mathcal{W}$ is "adic" and the operators F_m, $m \in \mathbb{N}^+$, extends to endomorphisms F_m for

$$m \in \hat{\mathbb{Z}}/_{\hat{\mathbb{Z}}^*} = \prod_p p^{N \cup \{\infty\}} \equiv \mathcal{CRing}(\mathcal{W}, \mathbb{Z}) \tag{0.49}$$

the super-natural-numbers, interpolating between the rank 1 projection $F_0(\phi_n) \equiv \varphi(n) \cdot \phi_1$, and the identify $F_1 = \mathrm{Id}_{\mathcal{W}}$:

$$\begin{aligned} \mathrm{tr}(\phi_n) = t_1(\phi_n) = \mu(n) \quad & \text{the Möbuis function} \\ \mathrm{tr}(F_0\phi_n) = t_0(\phi_n) = \varphi(n) \ & \text{the Euler function} \end{aligned} \tag{0.50}$$

We can complexify,

$$\mathcal{W}_{\mathbb{C}} = \mathbb{C} \otimes \mathcal{W} = \oplus_n \mathbb{C} \cdot \phi_n$$

and complete $\mathcal{W}_{\mathbb{C}}$ with respect to the Hermitian form

$$\langle f, g \rangle := \text{coefficient of } \phi_1 \text{ in the product } f \cdot \overline{g} \tag{0.51}$$

obtaining a Hilbert space

$$\mathcal{H} = \widehat{\mathcal{W}}_{\mathbb{C}} = \widehat{\bigoplus_n} \mathbb{C}\phi_n$$

with orthogonal basis ϕ_n,

$$\|\phi_n\|^2 = \varphi(n)$$

The "Fourier-transform" gives an isomorphism

$$\begin{aligned} \mathcal{H} &\xrightarrow{\ \sim\ } L_2\left(\hat{\mathbb{Z}}, dm\right)^{\hat{\mathbb{Z}}^*}, \quad \text{dm-additive Haar probability measure} \\ f &\longmapsto \hat{f}(m) := \mathrm{tr}(F_m f) = t_1(F_m f) = t_m(f) \end{aligned} \tag{0.52}$$

Considering the "zeta operators"

$$\begin{aligned} \zeta(F, s) &= \sum_{n \geqslant 1} \frac{F_n}{n^s} = \prod_p \frac{1}{(1 - p^{-s} F_p)} \\ \zeta(F^*, t) &= \sum_{m \geqslant 1} \frac{F_m^*}{m^t} = \zeta(F, \overline{t})^* \end{aligned} \tag{0.53}$$

well defined on the dense subset $\mathcal{W}_{\mathbb{C}}$, and analytic for $\Re(s)$, $\Re(t) > 1$, we have Ramanujan's evaluations

$$t_1(\zeta(F,s)\phi_1) = \zeta(s) \text{ the Riemann's zeta function} \tag{0.54}$$

$$t_1(\zeta(F^*,t)\phi_1) = \frac{1}{\zeta(t)} \tag{0.55}$$

$$t_1(\zeta(F,s)\zeta(F^*,t)\phi_1) = \sum_{n,m\geqslant 1} \frac{C_n^m}{n^t m^s} = \frac{\zeta(s)}{\zeta(t)} \cdot \zeta(s+t-1) \tag{0.56}$$

Some Historical Remarks

It was Kurokawa, Ochiai, Wakayama [**KOW02**] who first suggested forgetting addition, so as to have interesting derivative in arithmetic, and worked only with the underlying multiplicative monoid. This approach was further developed by Deitmar [**Dei05**]. But from this perspective the "primes" (i.e. submonoids that are closed under multiplication by anything, and whose compliment is multiplicative closed) are arbitrary unions of prime ideals (and hence of the cardinality of the continuation for the ordinary primes of $\mathbb{Z}$).

In representation theory, Tits [**Tit57**] discovered the "Field with one element" $\mathbb{F}$, noticing that the convolution algebra over the finite field $\mathbb{F}_q$ with q elements

$$\mathcal{H}_q = \mathbb{C}\left[B(\mathbb{F}_q)\Big\backslash \mathrm{GL}_n(\mathbb{F}_q)\Big/ B(\mathbb{F}_q)\right]$$

with B the Borel subgroup of upper triangular matrices in GL_n, has a fixed set of generators, and relations that depends on the parameter q, and setting $q = 1$ one obtains

$$\mathcal{H}_q\Big|_{q=1} = \mathbb{C}\left[S_n\right]$$

the group ring of the symmetric group $S_n = GL_n(\mathbb{F})$, the Weyl group of GL_n (and similarly for more general algebraic groups). This motivated Soule [**Sou08**] to try and define the collection of varieties over $\mathbb{F}$ as a sub-collection of the varieties over $\mathbb{Z}$ (obtained by "base change" from the hypothetical $\mathbb{F}$ to $\mathbb{Z}$), mainly toric varieties. This was generalized by Toën and Vaquie [**TV05**] who used relative algebraic geometry to construct schemes below spec $\mathbb{Z}$.

After hearing a lecture by Soule I was motivated to develop algebraic geometry using "generalized rings", associating with an ordinary ring the collection of matrices over it [**Har07**], but unfortunately used both the operations of block-direct-sum and of tensor products — thus I restricted attention to the "totally commutative" props, and so the arithmetical surface reduced to its diagonal $\mathbb{Z} \otimes_{\mathbb{F}} \mathbb{Z} = \mathbb{Z}$. In [**Har17b**] I corrected this, and dropped the tensor product from the structure, but unfortunately kept the involution given by taking the transpose of a matrix (or of identifying the operads $\mathcal{P}^{-}$ and $\mathcal{P}^{+}$ for bios). Keeping the involution complicated the development of the spectrum associated with a commutative prop or bio. The only advantage of having the involution is to fix the l_2-norm at the real and complex integers. If we do not insist on having involution we can use in (0.7) or (0.9) the l_p-norms, $p \in [1, \infty]$,

$$\|x_1, \ldots, x_m\|_p := (|x_1|^p + \cdots + |x_m|^p)^{1/p}$$

$$\|x_1, \ldots, x_m\|_{\ell_\infty} := max\{|x_1|, \ldots, |x_m|\}$$

and we obtain the ℓ_p-version of the real and complex integers $\mathbb{Z}_{\mathbb{R}}\{1/p\}$ and $\mathbb{Z}_{\mathbb{C}}\{1/p\}$, with

$$\mathbb{Z}_{\mathbb{R}}\{\sigma\}^{\mathrm{op}} = \mathbb{Z}_{R}\{1-\sigma\}, \quad \mathbb{Z}_{\mathbb{C}}^{\mathrm{op}} = \mathbb{Z}_{\mathbb{C}}\{1-\sigma\}$$

$$\mathbb{Z}_{\mathbb{R}}\left\{\frac{1}{2}\right\} \equiv \mathbb{Z}_{\mathbb{R}}, \quad \mathbb{Z}_{\mathbb{C}}\left\{\frac{1}{2}\right\} \equiv \mathbb{Z}_{\mathbb{C}}$$

There were some other approaches to "$\mathbb{F}_1$-geometry" using "Blueprints" by Lorscheid [**Lor16**]; using Λ-ring structure as descent data from $\mathbb{Z}$ to $\mathbb{F}$ by Borger [**Bor09**]; or using monads by Durov [**Dur08**]. In all these approaches the arithmetical surface reduces to its diagonal, $\mathbb{Z} \otimes \mathbb{Z} = \mathbb{Z}$, and moreover one often gets wrong answer at the "real prime" (e.g. in Durov's approach $\mathrm{GL}_n(\mathbb{Z}_{\mathbb{R}})$ is the finite group of symmetries of the ℓ_1-unit ball, rather than the infinite orthogonal group $O(n)$, the symmetries of the ℓ_2-unit ball). See [**PL11**] for a review of some of the interconnections between these approaches.

The idea of correcting Hochschild homology so that it will work correctly also in positive characteristic by "changing the initial object of algebra, the integers $\mathbb{Z}$, to the initial object of stable homotopy $\mathbb{S}$, the sphere spectrum", goes back to Goodwillie, Waldhausen, and to Bökstedt, who constructed Topological Hochschild homology. This approach to $\mathbb{F}$, as derived algebra, was developed by Connes and Consani [**CC16**], (their earlier approach to $\mathbb{F}$ was an elaboration of Soule's approach). They used Segal's approach to connective spectra via Γ-sets, and is nowadays part of the theory of derived algebra and geometry as developed by Lurie using the language of infinity categories, or the "quasi-categories" of Boardman and Vogt [**BV73**] and André Joyal [**Joy02**].

We end this introduction by giving a short summary of the content of the various chapters.

In Chapter 1, we give the definition of $\mathcal{P}rops$, and their commutativity. In Chapter 2, we describe the (full and faithful) embeddings of $\mathcal{R}ings$ in $\mathcal{P}rops$. In Chapter 3, we give the definition of $\mathcal{B}ios$, Chapter 4, discuss their commutativity. In Chapter 5, we describe the ideals and primes of a commutative $\mathcal{B}io$, A, and in Chapter 6, we describe the (compact, sober) Zariski topology on $\mathit{spec}(A)$. In Chapter 7, we give the basic facts for localizations, and in Chapter 8, construct the sheaf of $\mathcal{B}io/\mathcal{P}rops$ over $\mathit{spec}(A)$. In Chapter 9, we describe the categories of $\mathcal{B}io/\mathcal{P}rops$ schemes, and in Chapter 10, we describe the pro-categories of these, and the "compactification" $\overline{\mathit{spec}\mathcal{O}_K}$, K a number field. In Chapter 11, we define valuation props and bios, we show that they correspond in the rank 1 case to real valued valuations; for a number field K they correspond to the finite primes $\mathfrak{p} \subseteq \mathcal{O}_K$ or to the real/complex "primes" $\sigma : K \hookrightarrow \mathbb{C}$, $\sigma \sim \overline{\sigma}$. We also show how the operad structure is compatible with Beta-integrals, and give the "zeta machine": given a compact homogeneous valuation prop B, together with a map $\mathbb{N} \to K = (B(1))\backslash\{0\})^{-1}(B) \equiv$ fraction field of B, the zeta machine produces a meromophic function $L(B, s)$, normalized by $L(B, 1) = 1$:

$$L(\mathbb{Z}_p, s) = \frac{\zeta_p(s)}{\zeta_p(1)}, \text{ with}$$

$$\zeta_p(s) = (1 - p^{-s})^{-1}, \quad \zeta_{\mathbb{R}}(s) = 2^{\frac{s}{2}}\Gamma\left(\frac{s}{2}\right), \quad \zeta_{\mathbb{C}}(s) = \Gamma(s)$$

In Chapter 12, we describe for a $\mathcal{B}io/\mathcal{P}rop$ A, and a category $\mathcal{C}$, the A-objects in $\mathcal{C}$. In Chapter 13, we sin again with addition, taking $\mathcal{C} = \mathcal{A}b$ we have the category of A-modules, isomorphic to a category of abelian group objects in the category of $\mathcal{B}io/\mathcal{P}rop$ over

A giving rise to derivations, and to the Kahler differentials representing them. These all follow Quillen's approach and arise from our attempt to linearize geometry infinitesimally. We describe explicitly the $\mathbb{Z}$-prop-module of Kahler differentials $\Omega(\mathbb{Z}/\mathbb{F})$. In Chapter 14, we bring in the simplicial $\mathcal{Bio}/\mathcal{Prop}$, and the Quillen model structure on them, giving rise to the cotangent complex of a homomorphism of $\mathcal{Bio}/\mathcal{Prop}$. In Chapter 15, we define briefly the basic properties of maps of $\mathcal{Bio}/\mathcal{Prop}$ schemes. In Chapter 16, we give for a $\mathcal{Prop}$ A, commutative or not, the infinite loop space $\mathbb{K}(A)$ whose homotopy groups are the higher K-groups of A; we do it by a very concrete version of Quillen's $S^{-1}S$ construction. In Chapter 17, we give the Witt ring story, and the basic example of the special λ-ring $\mathcal{W} = \mathcal{W}(\overline{\mathit{spec}\mathbb{Z}})$. In Chapter 18, we give the (close) symmetric monoidal structure on (totally-commutative) A-sets for a $\mathcal{Bio}$ A. We briefly describe the stabilization of simplicial A-sets using "symmetric spectra" or modules over the sphere spectrum — the point being is (Jeff Smith idea [**HSS00**]) that the spheres $S^{\cdot} = \{S^n\}_{n\geqslant 0} \in (\mathcal{SSet})^{\amalg_n S_n} \equiv \sum(\mathbb{F})$ are a **commutative** monoid. This shows very clearly the interaction of homotopy and arithmetic via $\mathbb{F} \subseteq \mathbb{Z}$. Extending scalars to an arbitrary prop scheme X we get the sheaf of commutative monoids symmetric spectra S^*_X over X, it is the topological Hochschild homology.

This book is not a definitive treatment of non additive geometry, it rather leaves a lot of open ends, and new allies for the reader to go on exploring. For simplicity and clarity we do algebraic geometry in Grothendieck's EGA style, so schemes are obtained by glueing spectra using the Zariski topology, rather than the more modern approach of viewing them as "functors of points" which are locally representable with respect to some Grothendieck topology (although props and bios have interesting topologies). We try to avoid most technicalities, proofs or calculations, and we concentrate on the concepts and ideas. Two exceptions are the calculation of the global sections of the sheaf associated with a prop or a bio (Theorem 8.1), and the proof of Ostrowski theorem- the calculation of the valuations on a number field (Theorem 11.2). We included these so that the reader can see that they are exactly the same as the proofs of the analogue classical statements only formulated in the language of props or bios (the reader may skip these proofs on a first reading). For the same reason we avoid the globalizations of Chapters 16 and

18, which are best done in the more technical language of infinite categories. We use the old excuse of the giants:

> "I have tried to avoid numerical computations, thereby following Riemann's postulate that proofs should be given through ideas not voluminous computations".
>
> *David Hilbert*

Chapter 1

Props

Definition 1.1. A **prop** P is a strict-symmetric-monoidal category that is generated by one object x.

Thus we have sets $P_{n,m} = P(x^{\oplus m}, x^{\oplus n})$ with an action of $S_n \times S_m^{\mathrm{op}}$ (i.e. S_n acting on the left, S_m acting on the right, and these actions commute) and composition

$$\begin{aligned} P_{n,m} \times P_{m,\ell} &\longrightarrow P_{n,\ell} \\ f, g &\longmapsto f \circ g \end{aligned} \tag{1.1}$$

that is S_m-invariant, $S_n \times S_\ell^{\mathrm{op}}$-covariant, associative

$$(f \circ g) \circ h = f \circ (g \circ h) \tag{1.2}$$

and unital

$$1_n \circ f = f = f \circ 1_m, \; f \in P_{n,m}, \; 1_n \in P_{n,n} \tag{1.3}$$

We also assume $P_{0,m} = \{0\}$, $P_{n,0} = \{0\}$, and $0 := x^0$, the unit object of the monoidal structure, is an initial and final object of P.

The monoidal structure is given by

$$\begin{aligned} P_{n_1,m_1} \times P_{n_2,m_2} &\longrightarrow P_{n_1+n_2,m_1+m_2} \\ f_1, f_2 &\longmapsto f_1 \oplus f_2 \end{aligned} \tag{1.4}$$

It is $S_{n_1} \times S_{n_2} \subseteq S_{n_1+n_2}$ and $S_{m_1} \times S_{m_2} \subseteq S_{m_1+m_2}$ covariant, functorial

$$\begin{gathered}(f_1 \oplus f_2) \circ (g_1 \oplus g_2) = (f_1 \circ g_1) \oplus (f_2 \circ g_2) \\ 1_{n+m} = 1_n \oplus 1_m\end{gathered} \tag{1.5}$$

strict monoidal

$$\begin{gathered}(f_1 \oplus f_2) \oplus f_3 = f_1 \oplus (f_2 \oplus f_3) \\ f \oplus \mathrm{id}_0 = f = \mathrm{id}_0 \oplus f\end{gathered} \tag{1.6}$$

and symmetric

$$f_2 \oplus f_1 = \tau_{n_1,n_2} \circ (f_1 \oplus f_2) \circ \tau_{m_2,m_1} \tag{1.7}$$

with $\tau_{n,m} \in S_{n+m}$ given by

$$\tau_{n,m}(i) = \begin{cases} n+i & i \leqslant m \\ i-m & i > m \end{cases} \tag{1.8}$$

pictorially $\tau_{n,m}$:

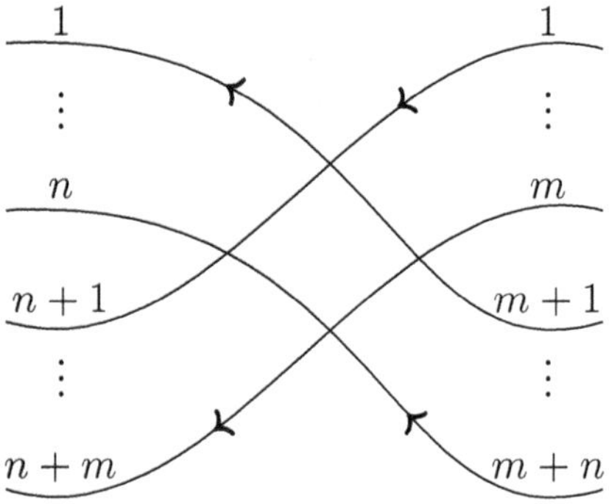

(1.9)

For props $\mathcal{P}$ and $\mathcal{Q}$ we put

$$\mathcal{P}rop(\mathcal{P}, \mathcal{Q}) = \left\{ \begin{array}{l} f_{n,m} \in \mathrm{Set}(\mathcal{P}_{n,m}, \mathcal{Q}_{n,m}), \quad f(\sigma \cdot p \cdot \sigma') = \sigma \cdot f(p) \cdot \sigma', \\ f(p_1 \circ p_2) = f(p_1) \circ f(p_2), f(1) = 1, \\ f(p_1 \oplus p_2) = f(p_1) \oplus f(p_2) \end{array} \right\} \tag{1.10}$$

so that we have a category $\mathcal{P}rop$.

Remark 1.1. Given a strict symmetric monoidal category $\mathcal{E}$, where the unit is an initial and final object, and an object $x \in \mathcal{E}^0$ we have the prop $\mathrm{End}_{\mathcal{E}}(x)$ with

$$\mathrm{End}_{\mathcal{E}}(x)_{n,m} := \mathcal{E}(x^{\oplus m}, x^{\oplus n}) \tag{1.11}$$

Remark 1.2. Replacing everywhere $\mathcal{S}et$ by $\mathcal{E}$ we obtain the category of props in $\mathcal{E}$: $\mathcal{P}rop(\mathcal{E})$. Thus we have the topological props $\mathcal{P}rop(\mathcal{T}op)$ and the simplicial props $\mathcal{P}rop(\mathcal{S}et^{\Delta^{\mathrm{op}}}) \equiv \mathcal{P}rop^{\Delta^{\mathrm{op}}}$.

Remark 1.3. The category $\mathcal{P}rop$ has an involution $P \mapsto P^{\mathrm{op}}$ with

$$P^{\mathrm{op}}_{n,m} := P_{m,n} \tag{1.12}$$

We have therefore the props with an involution $f \mapsto f^t$, $P_{n,m} \xrightarrow{\sim} P_{m,n}$ satisfying

$$f^{tt} = f, \quad (f_1 \circ f_2)^t = f_2^t \circ f_1^t, \quad 1^t = 1, \quad (f_1 \oplus f_2)^t = f_1^t \oplus f_2^t \tag{1.13}$$

and the category $\mathcal{P}rop^t$ (with maps preserving the involutions).

Definition 1.2. A prop P will be called **central** if the monoid $P_{1,1}$ is commutative and **central**:

$$a \cdot p := \left(\bigoplus^{n} a\right) \circ p \equiv p \circ \left(\bigoplus^{m} a\right) \quad \text{for } a \in P_{1,1},\ p \in P_{n,m} \tag{1.14}$$

A prop P will be called **commutative** if it is central and we have

$$(b \circ d) \cdot p \equiv \left(\bigoplus^{n} b\right) \circ \sigma_{n,k} \circ \left(\bigoplus^{k} p\right) \circ \sigma_{k,m} \circ \left(\bigoplus^{m} d\right) \tag{1.15}$$

for $b \in P_{1,k}$, $p \in P_{n,m}$, $d \in P_{k,1}$.

Remark 1.4. Given $p_j \in P_{n,m}$ for $j = 1, \ldots, k$, and given $b \in P_{1,k}$, $d \in P_{k,1}$, we have the "linear combination":

$$\left(\bigoplus^{n} b\right) \circ \sigma_{n,k} \circ \left(\bigoplus_{j=1}^{k} p_j\right) \circ \sigma_{k,m} \circ \left(\bigoplus^{m} d\right) \quad \in P_{n,m}$$

Indeed, if $\mathcal{P}$ is the prop of all matrices over a ring then this expression is equal to the linear combination

$$\sum_{j=1}^{k} b_j \cdot p_j \cdot d_j, \quad b = (b_1, \ldots, b_k), \quad d = \begin{pmatrix} d_1 \\ \vdots \\ d_k \end{pmatrix}$$

Thus commutativity means that the linear combination of p with itself is equal to p times the "scalar" $b \circ d \in \mathcal{P}_{1,1}$.

A prop $\mathcal{P}$ will be called **totally-commuative** if we have for $p \in \mathcal{P}_{p_0,p_1}$, $q \in \mathcal{P}_{q_0,q_1}$,

$$\begin{aligned} p \otimes q &:= \left(\bigoplus^{q_0} p\right) \circ \sigma_{q_0,p_1} \circ \left(\bigoplus^{p_1} q\right) \\ &\equiv \sigma_{q_0,p_0} \circ \left(\bigoplus^{p_0} q\right) \circ \sigma_{p_0,q_1} \circ \left(\bigoplus^{q_1} p\right) \circ \sigma_{q_1,p_1} \end{aligned} \tag{1.16}$$

Here $\sigma_{m,n} \in S_{m \cdot n}$ is the permutation

$$(i-1) \cdot n + j = \sigma_{m,n}\big((j-1) \cdot m + i\big)$$

see (0.25) for its diagram.

Remark 1.5. Note that total-commutativity implies commutativity.

It will be convenient to have the intermediate category of "**fully-commutative**" props which are commutative, (1.15), and also satisfy (1.16) when $p_1 = q_0 = 1$ or when $p_0 = q_1 = 1$. These are identity (0.23), or pictorially the identification of (0.24) with (0.25).

We let

$$\begin{gathered} \mathcal{P}^T \hookleftarrow \mathcal{P}, \quad \mathcal{P}^f \hookleftarrow \mathcal{P}, \quad \mathcal{P}^C \hookleftarrow \mathcal{P} \\ \mathcal{C}_T\mathcal{P}rop \subseteq \mathcal{C}_f\mathcal{P}rop \subseteq \mathcal{CP}rop \subseteq \mathcal{P}rop \end{gathered} \tag{1.17}$$

denote the full subcategory of (respectively totally-/fully-) commutative props, and we let $\mathcal{P} \twoheadrightarrow \mathcal{P}^C$ (respectively $\mathcal{P} \twoheadrightarrow \mathcal{P}^T$ $\mathcal{P} \twoheadrightarrow \mathcal{P}^f$) denote the maximal (totally-/fully) commutative quotient of $\mathcal{P}$ giving the left adjoints of the inclusions (1.17).

All these categories are complete and co-complete. All limits are formed in Sets (as the sub-prop of coherent sequences in the product prop). Moreover, directed co-limits are also formed in sets (as the quotient of the disjoint sum modulo an equivalence relation).

The initial object of these categories is the "**field with one element**" $\mathbb{F}$, of (0.19). The final object is the zero prop where all the sets $P_{n,m}$ reduce to a singleton (or equivalently, $0 = 1$ in $P_{1,1}$).

Chapter 2

Associating Props with Rings

We do not require the underlying additive structure of our rings to be a group, thus the definition of Rings can be generalized to "Rigs".

Definition 2.1. A **Rig** ($\equiv$ Ring without Negatives) is a set R with two operations of addition and multiplication

$$R \times R \rightrightarrows R, \quad (x, y) \longmapsto \begin{matrix} x+y \\ x \cdot y \end{matrix} \tag{2.1}$$

both associative and unital

$$\begin{aligned} &(x+y)+z = x+(y+z), \quad (x \cdot y) \cdot z = x \cdot (y \cdot z) \\ &x+0 = x = 0+x, \quad x \cdot 1 = x = 1 \cdot x \end{aligned} \tag{2.2}$$

with addition always commutative $x + y = y + x$, and distributive

$$\begin{gathered} (x_1 + x_2) \cdot y = (x_1 \cdot y) + (x_2 \cdot y), \; x \cdot (y_1 + y_2) = (x \cdot y_1) + (x \cdot y_2) \\ \text{and } x \cdot 0 = 0 = 0 \cdot x \end{gathered} \tag{2.3}$$

A rig with involution $R \in \mathcal{R}ig^t$ is a rig R with an involution

$$R \xrightarrow{\sim} R^{\mathrm{op}}, \quad x \mapsto x^t$$

satisfying

$$\begin{aligned} &x^{tt} = x, \quad 0^t = 0, \quad 1^t = 1 \\ &(x+y)^t = x^t + y^t, \quad (x \cdot y)^t = y^t \cdot x^t \end{aligned} \tag{2.4}$$

A rig is commutative $R \in \mathcal{CRig}$ if multiplication is commutative,

$$x \cdot y = y \cdot x$$

A map of rigs $\varphi : R \to R'$ is a set map preserving the operations and constants

$$\begin{aligned} &\varphi(0) = 0, \ \varphi(1) = 1, \quad \varphi(x+y) = \varphi(x) + \varphi(y), \\ &\varphi(x \cdot y) = \varphi(x) \cdot \varphi(y) \end{aligned} \tag{2.5}$$

In the self-dual case $\mathcal{Rig}^t$, φ should also preserve the involution,

$$\varphi(x^t) = \varphi(x)^t$$

Thus we have categories and functors, with U the functor that forget the involution

$$\begin{array}{ccc} \mathcal{Rig}^t & \xrightarrow{\ U\ } & \mathcal{Rig} \\ & \nwarrow & \cup\!\! \mid \\ & & \mathcal{CRig} \end{array} \tag{2.6}$$

(for a commutative rig the identity is an involution).

There is a similar diagram with $\mathcal{Ring}$ instead of $\mathcal{Rig}$.

For example, We have the commutative rigs

$$\mathscr{B} = \{0,1\} \hookrightarrow \mathscr{I} = [0,1] \hookrightarrow \mathscr{R}_\infty = [0,\infty] \tag{2.7}$$

with the usual multiplication $x \cdot y$, and with the "tropical" addition

$$x +_\infty y : \overset{def}{=} max\{x,y\}. \tag{2.8}$$

The "valuation" rig $\mathscr{I} = [0,1]$, and its residue field, the Boolean rig $\mathscr{B} = \{0,1\}$, are basic for "tropical geometry".

Using the "ℓ_p-addition", $p \in [1,\infty]$,

$$x +_p y = (x^p + y^p)^{1/p} \tag{2.9}$$

we get the rigs $\mathscr{R}_p = [0,\infty]$ interpolating between $\mathscr{R}_1$, ordinary addition, to $\mathscr{R}_\infty$, tropical addition, with $f^p : \mathscr{R}_p \xrightarrow{\sim} \mathscr{R}_1$, $f^p(x) \equiv x^p$, $p < \infty$.

The inclusion of abelian groups in commutative monoids

$$\begin{array}{c} \mathcal{CMon} \\ K \left(\downarrow \;\; \uparrow \right) \\ \mathcal{Ab} \end{array} \tag{2.10}$$

has the left adjoint K-functor, Grothendieck's localization of addition.

For $M \in \mathcal{CMon}$, the group $K(M)$ consists of formal differences $m_1 - m_2$, $m_i \in M$, with

$$\begin{aligned} & m_1 - m_2 = m_1' - m_2' \iff \exists m_0 \in M \\ & m_0 + m_1 + m_2' = m_0 + m_1' + m_2 \end{aligned} \tag{2.11}$$

It gives the left adjoints

$$\begin{array}{ccccc} \mathcal{CRig} & \hookrightarrow & \mathcal{Rig} & \xrightarrow{f} & \mathcal{CMon} \\ K \left(\;\; \cup\!\shortmid \right. & & K \left(\;\; \cup\!\shortmid \right. & & K \left(\;\; \cup\!\shortmid \right. \\ \mathcal{CRing} & \hookrightarrow & \mathcal{Ring} & \xrightarrow{f} & \mathcal{Ab} \end{array} \tag{2.12}$$

with f the functor forgetting multiplication.

For a rig A we associate the prop (denoted by the same letter) A, with $A_{n,m}$ the n by m matrices with entries in A, $n, m \geqslant 0$. These form a category with respect to matrix multiplication, and is strict-symmetric-monoidal with respect to block-direct-sum of matrices.

For $n = 0$ (respectively, $m = 0$) we have $A_{0,m} = \{\phi_m\}$ (respectively, $A_{n,0} = \{\phi_m^t\}$) the empty row (respectively, column) of length m (respectively, n), and in particular $A_{0,0} = \{\phi_0 = \mathrm{Id}_0\}$ is the empty matrix so $a \oplus \mathrm{Id}_0 = a = \mathrm{Id}_0 \oplus a$. Note that $a \oplus \phi_m$ (respectively, $a \oplus \phi_m^t$) is obtained from a by adding to it m columns (respectively, rows) of zeros. Thus with $1 \in A_{1,1}$, using $\oplus$, and the actions of the symmetric groups permuting the rows and columns, we can obtain any element of $\mathbb{F}$.

Remark 2.1. Note that when the rig A is commutative, the associated prop A is totally-commutative (and $p \otimes q$ is just the tensor-product of the matrices p and q).

Example 2.1. For $p \in [1, \infty]$, let $\mathbb{R}_{\ell_p}$ (respectively, $\mathbb{C}_{\ell_p}$) denote the sub-prop of $\mathbb{R}$ (respectively, $\mathbb{C}$), with $(\mathbb{R}_{\ell_p})_{n,m}$ (respectively, $(\mathbb{C}_{\ell_p})_{n,m}$) all the matrices taking the unit ℓ_p m-ball to the unit ℓ_p n-ball,

$$\begin{aligned}(\mathbb{R}_{\ell_p})_{n,m} &= \left\{a \in \mathbb{R}_{n,m}, \|a \circ x\|_{\ell_p} \leqslant \|x\|_{\ell_p} \text{ all } x \in \mathbb{R}_{m,1}\right\} \\ \|x_1, x_2, \ldots, x_n\|_{\ell_p} &= (|x_1|^p + |x_2|^p + \cdots + |x_n|^p)^{1/p} \\ \|x_1, x_2, \ldots, x_n\|_{\ell_\infty} &= max\,\{|x_1|, |x_2|, \ldots, |x_n|\}\end{aligned} \tag{2.13}$$

For $p = 2$, we get the real (respectively, complex) "integers" (0.6) (respectively, (0.8))

$$\mathbb{R}_{\ell_2} \equiv \mathbb{R}_{\mathbb{Z}}, \quad \mathbb{C}_{\ell_2} \equiv \mathbb{C}_{\mathbb{Z}} \tag{2.14}$$

We have

$$\left(\mathbb{R}_{\ell_p}\right)^{\mathrm{op}} = \mathbb{R}_{\ell_q}, \quad \left(\mathbb{C}_{\ell_p}\right)^{\mathrm{op}} = \mathbb{C}_{\ell_q} \tag{2.15}$$

where

$$1/p + 1/q = 1 \tag{2.16}$$

thus the "integers" are special in that they have an involution.

Example 2.2. For the initial object of $\mathcal{R}ig$ (respectively, $\mathcal{R}ing$) we have the prop $\mathbb{N}$ (respectively, $\mathbb{Z}$). Having the elements

$$v = (1, 1) \in \mathbb{N}_{1,2}, \quad v^t = \begin{pmatrix}1\\1\end{pmatrix} \in \mathbb{N}_{2,1} \tag{2.17}$$

we have addition: for $a, b \in \mathbb{N}_{n,m}$

$$a + b = \left(\bigoplus^{n} v\right) \circ \sigma_{n,2} \circ (a \oplus b) \circ \sigma_{2,m} \circ \left(\bigoplus^{m} v^t\right) \tag{2.18}$$

By adding elements of $\mathbb{F}$ we can obtain any matrix over $\mathbb{N}$, so v, v^t generate $\mathbb{N}$ as a prop. We have the relations:

Associativity:

$$v \circ (v, 1) = v \circ (1, v) (= (1, 1, 1)) \tag{2.19}$$

graphically,

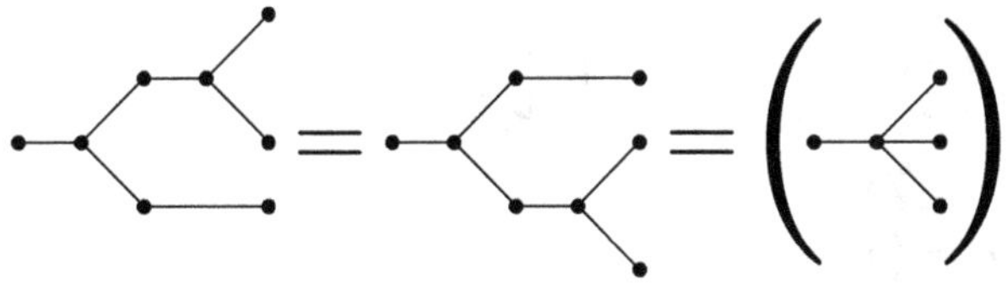

and the transposed relation

$$\left(\begin{pmatrix}1\\1\\1\end{pmatrix} = \right) \qquad (1, v^t) \circ v^t = (v^t, 1) \circ v^t \tag{2.20}$$

Commutativity:

$$v \circ \begin{pmatrix}0 & 1\\1 & 0\end{pmatrix} = v \qquad v^t = \begin{pmatrix}0 & 1\\1 & 0\end{pmatrix} \circ v^t \tag{2.21}$$

Unit

$$v \circ \begin{pmatrix}1\\0\end{pmatrix} = 1 \qquad 1 = (1, 0) \circ v^t \tag{2.22}$$

Similarly, $\mathbb{Z}$ is generated by v, v^t, and $(-1) \in \mathbb{Z}_{1,1}$, with the above relation for v, v^t and

$$(-1) \circ (-1) = 1 \tag{2.23}$$

Cancellation:

$$v \circ (1 \oplus (-1)) \circ v^t = 0 \tag{2.24}$$

In particular we have the **long vectors**

$$v_n = \underbrace{v \circ_1 v \circ_1 \cdots \circ_1 v}_{n-1} = (\underbrace{1, 1, \ldots, 1}_{n}) \tag{2.25}$$

and

$$v_n^t = v^t \,{}_1{\circ}\, v^t \,{}_1{\circ}\, v^t \,{}_1{\circ} \cdots {}_1{\circ}\, v^t = \left.\begin{pmatrix}1\\ \vdots\\ 1\end{pmatrix}\right\} n$$

Any $a = (a_{i,j}) \in \mathbb{N}_{n,m}$ can be brought (using the above relation, and $(n) \equiv v_n \circ v_n^t$) into a canonical form as

$$a = \left(\bigoplus_{i=1}^{n} v_{a_{i,1}+\cdots+a_{i,m}} \right) \circ \sigma \circ \left(\bigoplus_{j=1}^{m} v^t_{a_{1,j}+\cdots+a_{n,j}} \right), \quad \sigma \in S_{\sum\limits_{i,j} a_{i,j}} \tag{2.26}$$

respectively, $a = (a_{i,j}) \in \mathbb{Z}_{n,m}$

$$\begin{aligned} a = \left(\bigoplus_{i=1}^{n} v_{|a_{i,1}|+\cdots+|a_{i,m}|} \right) \circ \sigma \circ \left(\bigoplus_{j=1}^{m} v^t_{|a_{1,j}|+\cdots+|a_{n,j}|} \right) \\ \sigma \in (\pm 1)^{\sum\limits_{i,j} |a_{i,j}|} \rtimes S_{\sum\limits_{i,j} |a_{i,j}|}. \end{aligned} \tag{2.27}$$

This shows the above relation generate all the relations.

Since $\mathbb{N}$, (respectively, $\mathbb{Z}$), is the initial object of $\mathcal{R}ig$ (respectively, $\mathcal{R}ing$), the full embedding of $\mathcal{R}ig$ into $\mathcal{P}rop$ gives an embedding into $\mathcal{P}rop$ object under $\mathbb{N}$, (respectively, $\mathbb{Z}$), and once we are under $\mathbb{N}$, we have the vectors v, v^t, and we have addition. These embeddings have the left adjoints functors R, given by $R(A)$ the Ri(n)g associated to the prop A,

$$\begin{aligned} R(A) &:= A_{1,1} \\ a_1 + a_2 &= v \circ (a_1 \oplus a_2) \circ v^t \\ a_1 \cdot a_2 &= a_1 \circ a_2 \end{aligned} \tag{2.28}$$

$$\begin{array}{ccc} \mathcal{R}ig & \overset{R}{\underset{}{\leftrightarrows}} & \mathcal{P}rop_{\mathbb{N}\backslash} \\ \text{IU} & & \text{IU} \\ \mathcal{R}ing & \overset{R}{\underset{}{\leftrightarrows}} & \mathcal{P}rop_{\mathbb{Z}\backslash} \end{array}$$

Chapter 3

Bio(perads)

Definition 3.1. A Bio (short for Bi-operad) is a pair $\mathcal{P} = (\mathcal{P}^-, \mathcal{P}^+)$ of closed symmetric operads $\mathcal{P}^-$ and $\mathcal{P}^+$ "acting" on each other.

Explicitly, we have sets $\mathcal{P}^-(n)$, $\mathcal{P}^+(n)$, $n \geqslant 0$, with an action of the symmetric group S_n on the right on $\mathcal{P}^-(n)$, and on the left on $\mathcal{P}^+(n)$,

$$\begin{array}{ccc} \mathcal{P}^-(n) \circlearrowright & S_n & \circlearrowleft \mathcal{P}^+(n) \\ & \Downarrow & \\ p \mapsto p \circ \sigma & \sigma & \sigma \circ q \leftarrow q \end{array} \tag{3.1}$$

In general we call the elements of $\mathcal{P}^-$ (respectively, $\mathcal{P}^+$) "operations" (respectively, "co-operations"), and usually denote them by p (respectively, q).

We assume $\mathcal{P}^\pm(0) = \{0^\pm\}$ are singleton, so that we are working with *closed* operads (one could also work with *open* operads, with $\mathcal{P}^\pm(0)$ empty).

We have **composition** maps

$$\begin{gathered} \mathcal{P}^-(n) \times \mathcal{P}^-(k_1) \times \cdots \times \mathcal{P}^-(k_n) \longrightarrow \mathcal{P}^-(k_1 + \cdots + k_n) \\ (p, p_1, \ldots, p_n) \mapsto p \circ (p_i) \\ \mathcal{P}^+(k_1 + \cdots + k_n) \longleftarrow \mathcal{P}^+(k_1) \times \cdots \times \mathcal{P}^+(k_n) \times \mathcal{P}^+(n) \\ (q_i) \circ q \leftarrow (q_1, \ldots, q_n, q) \end{gathered} \tag{3.2}$$

which are associative

$$(p \circ (p_i)) \circ (p_{ij}) = p \circ (p_i \circ (p_{ij})),$$

$$((q_{ij}) \circ q_i) \circ q = (q_{ij}) \circ ((q_i) \circ q) \tag{3.3}$$

and unital

$$\begin{aligned} 1^- \circ p = p \circ (1^-, \ldots, 1^-) = p, \quad 1^- \in \mathcal{P}^-(1), \\ \mathcal{P}^+(1) \ni 1^+, \quad q = (1^+, \ldots, 1^+) \circ q = q \circ 1^+ \end{aligned} \tag{3.4}$$

and "$(S_{k_1} \times \cdots \times S_{k_n}) \rtimes S_n \subseteq S_{k_1+\cdots+k_n}$-covariant".

We have also **action** maps

$$\begin{gathered} \mathcal{P}^-(k_1 + \cdots + k_n) \times \mathcal{P}^+(k_1) \times \cdots \times \mathcal{P}^+(k_n) \to \mathcal{P}^-(n) \\ (p, q_1, \ldots q_n) \mapsto p \overleftarrow{\circ} (q_i) \\ \mathcal{P}^+(n) \leftarrow \mathcal{P}^-(k_1) \times \cdots \times \mathcal{P}^-(k_n) \times \mathcal{P}^+(k_1 + \cdots + k_n) \\ (p_i) \overrightarrow{\circ} q \leftarrow\!\shortmid (p_1, \ldots, p_n, q) \end{gathered} \tag{3.5}$$

which are associative

$$\begin{aligned} p \overleftarrow{\circ} ((q_{ij} \circ q_i)) = (p \overleftarrow{\circ} (q_{ij})) \overleftarrow{\circ} (q_i) \\ (p_i) \overrightarrow{\circ} ((p_{ij}) \overrightarrow{\circ} q) = (p_i \circ (p_{ij})) \overrightarrow{\circ} q \end{aligned} \tag{3.6}$$

unital

$$p \overleftarrow{\circ} (1^-) = p, \quad q = (1^+) \overrightarrow{\circ} q \tag{3.7}$$

$S_{k_1} \times \cdots \times S_{k_n}$-invariant and S_n-covariant.

We also require **naturality** of the action maps, so we have

$$(p \circ (p_i)) \overleftarrow{\circ} (q_{ij}) = p \circ (p_i \overleftarrow{\circ} (q_{ij})) \tag{3.8}$$

pictorially

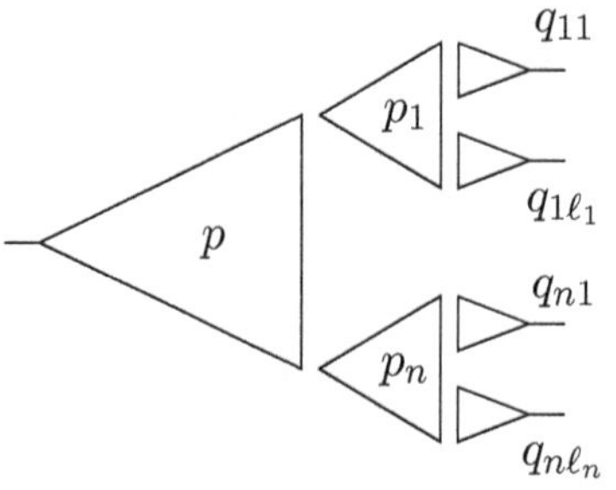

$$\begin{aligned} & p \in \mathcal{P}^-(n), \\ & p_i \in \mathcal{P}^-(k_{i1} + \cdots + k_{i,\ell_i}) \\ & q_{ij} \in \mathcal{P}^+(k_{ij}) \end{aligned}$$

and symmetrically,

$$((p_{ij}) \overrightarrow{\circ}\, q_i) \circ q = (p_{ij})\, \overrightarrow{\circ}\, ((q_i) \circ q) \tag{3.9}$$

We also require **co-naturality**, so

$$\big(p \circ (p_{ij})\big)\, \overleftarrow{\circ}\, (q_i) = p\, \overleftarrow{\circ}\, ((p_{ij}) \overrightarrow{\circ}\, q_i) \tag{3.10}$$

pictorially

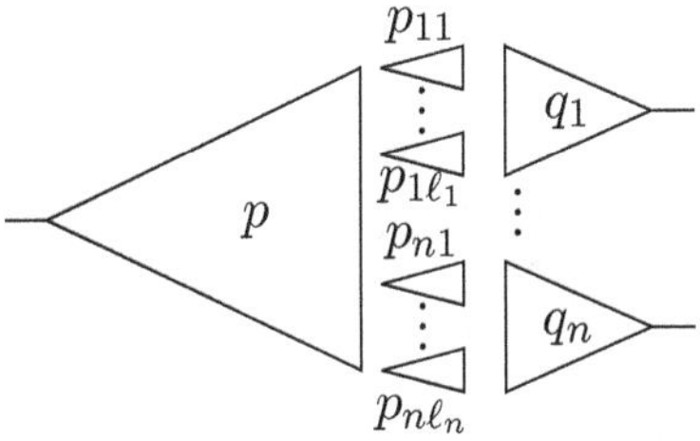

$$p \in \mathcal{P}^-(\ell_1 + \cdots + \ell_n)$$
$$p_{ij} \in \mathcal{P}^-(k_{ij})$$
$$q_i \in \mathcal{P}^+(k_{i1} + \cdots + k_{i\ell_i})$$

and symmmetrically,

$$(p_i\, \overleftarrow{\circ}\, (q_{ij}))\, \overrightarrow{\circ}\, q = (p_i) \overrightarrow{\circ}\, ((q_{ij}) \circ q) \tag{3.11}$$

Maps of bios $\varphi : \mathcal{P} \to \mathcal{Q}$ are pairs of maps of symmetric operads

$$\varphi^{\pm} : \mathcal{P}^{\pm} \to \mathcal{Q}^{\pm}$$

preserving the S_n-actions, the compositions and units, that also preserve the action maps:

$$\varphi^-\,(p\, \overleftarrow{\circ}\, (q_i)) = \varphi^-(p)\, \overleftarrow{\circ}\, (\varphi^+(q_i))$$
$$\varphi^+\,((p_i) \overrightarrow{\circ}\, q) = \big(\varphi^-(p_i)\big)\, \overrightarrow{\circ}\, \varphi^+(q)$$

Thus we have a category $\mathcal{B}io$.

Remark 3.1. The whole structure can be defined via the operations "at the jth entry", $1 \leqslant j \leqslant n$, obtained by adding 1's in all the non-relevant entries,

$$\circ_j : \mathcal{P}^-(n) \times \mathcal{P}^-(k) \longrightarrow \mathcal{P}^-(n+k-1)$$
$$\mathcal{P}^+(n+k-1) \longleftarrow \mathcal{P}^+(k) \times \mathcal{P}^+(n) \;:\; \circ_j$$
$$\overleftarrow{\circ_j} : \mathcal{P}^-(n+k-1) \times \mathcal{P}^+(k) \longrightarrow \mathcal{P}^-(n)$$
$$\mathcal{P}^+(n) \longleftarrow \mathcal{P}^-(k) \times \mathcal{P}^+(n+k-1) :\; \overrightarrow{\circ_j}$$

In this definition we took the operads $\mathcal{P}^{\pm}$ to be closed, so that we have also $\mathcal{P}^{-}(0) = \{0^{-}\}$ and $\mathcal{P}^{+}(0) = \{0^{+}\}$, and multiplication by $0^{\pm}$ at any place $j \in \{1, \ldots, n\}$ map $\mathcal{P}^{\pm}(n)$ to $\mathcal{P}^{\pm}(n-1)$, "erase an entry". Similarly $0^{\mp}$ act on $\mathcal{P}^{\pm}(n)$, at any place j, mapping it to $\mathcal{P}^{\pm}(n+1)$, "add a 0 entry".

Remark 3.2. The monoids $\mathcal{P}^{-}(1)$ and $\mathcal{P}^{+}(1)$ are canonically isomorphic via acting on the units

$$\begin{array}{ccc} \mathcal{P}^{-}(1) & \xleftrightarrow{\sim} & \mathcal{P}^{+}(1) \\ f^{-} & \longmapsto & f^{-} \overrightarrow{\circ} 1^{+} \\ 1^{-} \overleftarrow{\circ} f^{+} & \longleftarrow\!\shortmid & f^{+} \end{array}$$

We use this as an identification: It identifies 1^{-} with 1^{+}, and for $p \in \mathcal{P}^{-}(n)$, $q \in \mathcal{P}^{+}(n)$, $p \overrightarrow{\circ} q$ is identified with $p \overleftarrow{\circ} q$.

We will abuse notations and write $\circ$ for either $\overleftarrow{\circ}$ or $\overrightarrow{\circ}$.

Remark 3.3. Putting for $n \in \mathbb{Z}$,

$$\mathcal{P}\{n\} = \begin{cases} \mathcal{P}^{+}(n+1) & n \geqslant 0 \\ \mathcal{P}^{-}(1-n) & n \leqslant 0 \end{cases}$$

we have, $\mathcal{P} = \{0^{-}, 0^{+}\} \amalg \coprod_{n \in \mathbb{Z}} \mathcal{P}\{n\}$, and operations $o_j, \overleftarrow{o_j}, \overrightarrow{o_j}$ that map $\mathcal{P}\{n\} \times \mathcal{P}\{k\}$ to $\mathcal{P}\{n+k\}$. Thus for bios the grading by the "index" is more natural, but we will keep the operads notations.

Example 3.1. Given a strict symmetric monoidal category $\mathcal{E}$, where the unit is an initial and final object, and an object $X \in \mathcal{E}^{\circ}$, we have the bio $\mathrm{End}_{\mathcal{E}}(X)$ with

$$\mathrm{End}_{\mathcal{E}}^{-}(X)(n) = \mathcal{E}(X^{\otimes n}, X), \quad \mathrm{End}_{\mathcal{E}}^{+}(X)(n) \equiv \mathcal{E}(X, X^{\otimes n})$$

with all the compositions and actions given by the monoidal structure of $\mathcal{E}$.

Remark 3.4. Given a symmetric monoidal category $\mathcal{E}$, we can replace $\mathcal{S}et$ everywhere by $\mathcal{E}$ and obtain the category $\mathcal{B}io(\mathcal{E})$ of bios in $\mathcal{E}$. Thus we have the Topological bios, and the simplicial bios.

Remark 3.5. The category $\mathcal{B}io$ has an involution

$$\mathcal{P} = (\mathcal{P}^-, \mathcal{P}^+) \longmapsto \mathcal{P}^{op} = \left((\mathcal{P}^{op})^- \equiv \mathcal{P}^+, (\mathcal{P}^{op})^+ \equiv \mathcal{P}^-\right)$$

It takes the monoid $\mathcal{P}(1)$ to the opposite monoid $\mathcal{P}^{op}(1) \equiv \mathcal{P}(1)^{op}$.

We get the category $\mathcal{B}io^t$, with objects the bios $\mathcal{P}$ with an involution

$$p \longmapsto p^t : \mathcal{P} \xrightarrow{\sim} \mathcal{P}^{op}$$

and with maps preserving the involution. An object $\mathcal{P} \in \mathcal{B}io^t$ is completely described by the operad $\mathcal{P}^-$ and by the action of $\mathcal{P}^-$ on itself, the "contraction",

$$\begin{aligned} \mathcal{P}^-(n+k-1) \times \mathcal{P}^-(k) &\longrightarrow \mathcal{P}^-(n) \\ p,\ q &\longmapsto p \overleftarrow{\circ}_j q^t := p//_j q \end{aligned}$$

Thus the category $\mathcal{B}io^t$ is equivalent to the category of Generalized Rings of [**Har17b**]

Given a prop $\mathcal{P} = \{\mathcal{P}_{n,m}\}$ we get a bio $U\mathcal{P}$,

$$(U\mathcal{P})^-(n) := \mathcal{P}_{1,n}, \quad (U\mathcal{P})^+(n) := \mathcal{P}_{n,1} \tag{3.12}$$

with operations

$$p \circ (p_1 \cdots p_n) := p \circ (\bigoplus p_j), \quad (q_1, \ldots, q_n) \circ q := (\bigoplus q_j) \circ q \tag{3.13}$$

and actions

$$p \overleftarrow{\circ} (q_1, \ldots, q_n) := p \circ (\bigoplus q_j), \quad (p_1, \ldots, p_n) \overrightarrow{\circ} q := (\bigoplus p_j) \circ q \tag{3.14}$$

This gives a functor $U : \mathcal{P}rop \to \mathcal{B}io$. This functor has a left adjoint, taking a bio $B \in \mathcal{B}io$ to the free prop generated by B, $\mathcal{F}B \in \mathcal{P}rop$. The elements of $(\mathcal{F}B)_{n,m}$ can be described graphically, and non uniquely, as certain oriented graphs, with no loops, with m input edges, n output edges, and any inner vertex has either one output and n inputs (and is assigned an element of $\mathcal{P}^-(n)$), or has one input and n outputs (and is assigned an element of $\mathcal{P}^+(n)$).

For a rig A, the bio UA will be denoted simply by A, and it consists of all the row and column vectors over A. It will be convenient to view row vectors $(a_1, \ldots, a_n)$, (respectively, column vector $(b_1, \ldots, b_n)^t$), as (respectively, co-)operations from right to left:

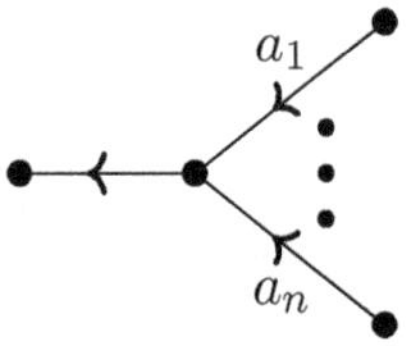

respectively,

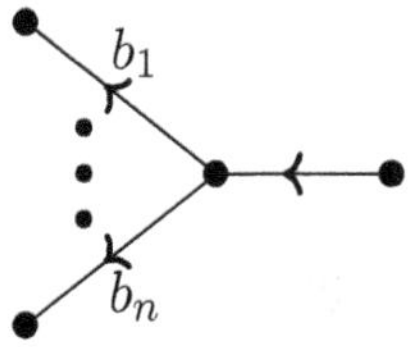

The operation of multiplication at the jth place given by

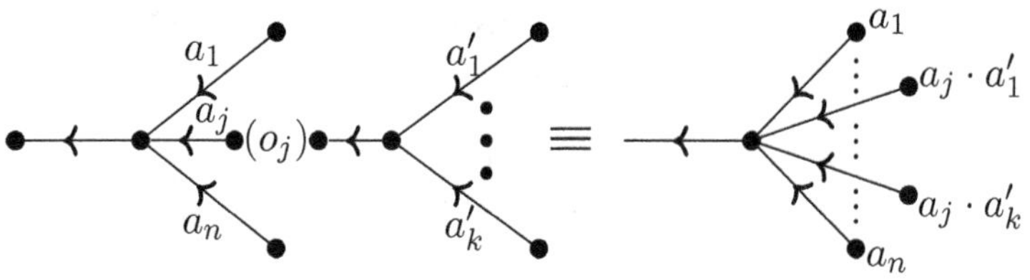

(3.15)

and similarly,

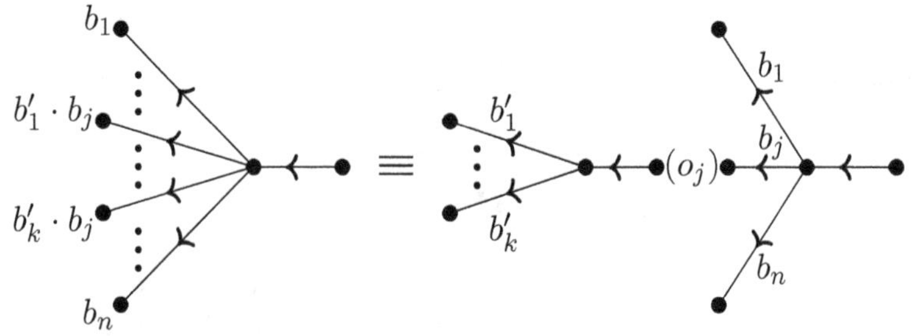

(3.16)

The actions are given by contractions

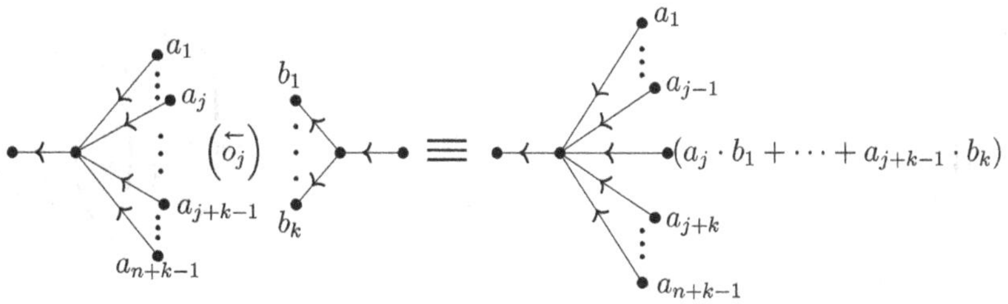

(3.17)

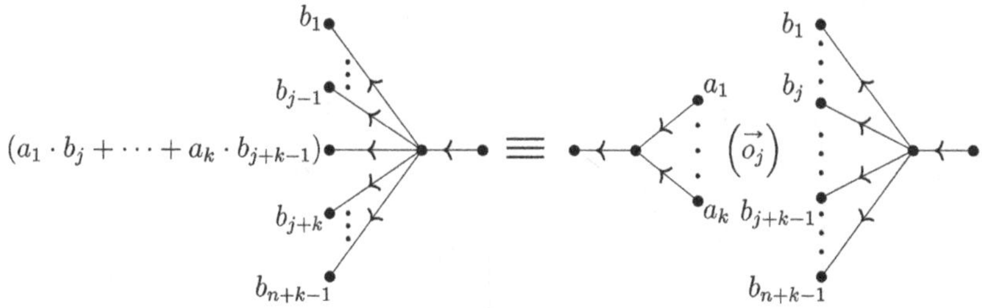

The diagram

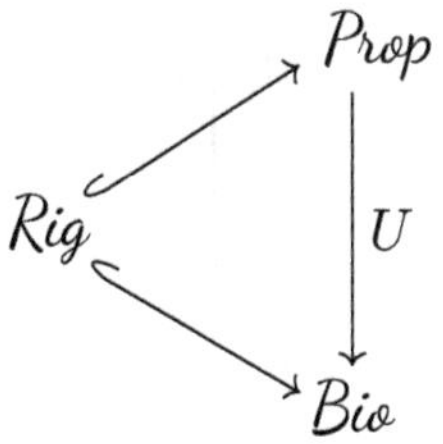

(3.18)

is commutative by definition. For a rig A, any matrix $(a_{ij}) \in A_{n,m}$ can be written as composition of rows and column vectors, so that the prop A is generated by the bio A:

$$\begin{pmatrix} a_{11} & \cdots & a_{1m} \\ \vdots & & \vdots \\ a_{n1} & \cdots & a_{nm} \end{pmatrix} \equiv \left(\bigoplus_{i=1}^{n} (a_{i,1}, \ldots, a_{i,m}) \right) \circ \sigma_{n,m} \circ \left(\bigoplus_{j=1}^{m} \begin{pmatrix} 1 \\ \vdots \\ 1 \end{pmatrix} \right)$$

$$\equiv \begin{pmatrix} a_{1,1} & \cdots & a_{1,m} & 0 & \cdots & 0 & \cdots & 0 & \cdots & 0 \\ 0 & \cdots & 0 & a_{2,1} & \cdots & a_{1,m_1} & 0 \cdots 0 & 0 & \cdots & 0 \\ & \vdots & & & \vdots & & \ddots & & \vdots & \\ 0 & \cdots & 0 & 0 & \cdots & 0 & \cdots & a_{n,1} & \cdots & a_{n,m} \end{pmatrix}$$

$$\times \begin{pmatrix} 1 & 0 & 0 & \cdots & 0 \\ 0 & 1 & 0 & \cdots & 0 \\ 0 & 0 & 1 & \cdots & 0 \\ \vdots & \cdots & & \ddots & 0 \\ 0 & \cdots & 0 & 0 & 1 \\ 1 & 0 & 0 & \cdots & 0 \\ 0 & 1 & 0 & \cdots & 0 \\ 0 & 0 & 1 & \cdots & 0 \\ \vdots & \cdots & & \ddots & 0 \\ 0 & \cdots & 0 & 0 & 1 \\ & & \vdots & & \\ 1 & 0 & 0 & \cdots & 0 \\ 0 & 1 & 0 & \cdots & 0 \\ 0 & 0 & 1 & \cdots & 0 \\ \vdots & \cdots & & \ddots & 0 \\ 0 & \cdots & 0 & 0 & 1 \end{pmatrix} \tag{3.19}$$

The initial element of $\mathcal{B}io$ is the field with one element $\mathbb{F} = U\mathbb{F}$,

$$\mathbb{F}^-(n) = \{0_n = (0, \ldots, 0, 0), \delta_1 = (1, \ldots, 0, 0), \delta_n = (0, \ldots, 0, 1)\}$$

$$\mathbb{F}^+(n) = \left\{ 0_n^t = \begin{pmatrix} 0 \\ \vdots \\ 0 \\ 0 \end{pmatrix}, \delta_1^t = \begin{pmatrix} 1 \\ \vdots \\ 0 \\ 0 \end{pmatrix}, \ldots, \delta_n^t = \begin{pmatrix} 0 \\ \vdots \\ 0 \\ 1 \end{pmatrix} \right\} \tag{3.20}$$

Example 3.2. The bio $\mathbb{N}$, (respectively, $\mathbb{Z}$), is again generated by

$$v = (1,1) \in \mathbb{N}^-(2) \text{ and } v^t = \begin{pmatrix} 1 \\ 1 \end{pmatrix} \in \mathbb{N}^+(2),$$

$$\text{respectively, and } (-1) \in \mathbb{Z}(1) \tag{3.21}$$

with the same relations as in Example 2.2, (2.19) until (2.24), written in the language of bios instead of props:

$$v \circ (v, 1) = v \circ (1, v), \quad (1, v^t) \circ v^t = (v^t, 1) \circ v^t \tag{3.22}$$

$$v \circ \begin{pmatrix} 0 & 1 \\ 1 & 0 \end{pmatrix} = v, \quad v^t = \begin{pmatrix} 0 & 1 \\ 1 & 0 \end{pmatrix} \circ v^t \tag{3.23}$$

$$v \circ \begin{pmatrix} 1 \\ 0 \end{pmatrix} = 1, \quad 1 = (1, 0) \circ v^t \tag{3.24}$$

respectively,

$$(-1) \circ (-1) = 1, \quad v \circ \big(1, (-1)\big) \circ v^t = 0 \tag{3.25}$$

We have the left adjoint functors R to the faithful embeddings of $\mathcal{R}i(n)g$ into $\mathcal{B}io$s under $\mathbb{N}$, (respectively, $\mathbb{Z}$),

$$\begin{array}{ccc} \mathcal{R}ig & \overset{R}{\underset{}{\leftrightarrows}} & \mathcal{B}io_{\mathbb{N}\backslash} \\ \text{IU} & & \text{IU} \\ \mathcal{R}ing & \overset{R}{\underset{}{\leftrightarrows}} & \mathcal{B}io_{\mathbb{Z}\backslash} \end{array} \tag{3.26}$$

$$R(A) := A(1), \quad a_1 + a_2 = v \circ (a_i) \circ v^t, \quad a_1 \cdot a_2 = a_1 \circ a_2$$

Example 3.3. For $p \in [1, \infty]$, we have the bios

$$U\mathbb{R}_{\ell_p} = (\mathbb{R}^-_{\ell_q}, \mathbb{R}^+_{\ell_p}), \quad \text{(respectively, } U\mathbb{C}_{\ell_p} = (\mathbb{C}^-_{\ell_q}, \mathbb{C}^+_{\ell_p}))$$

where

$$1/p + 1/q = 1, \quad \text{or} \quad q = (1 - p^{-1})^{-1} \tag{3.27}$$

and

$$\mathbb{R}^+_{\ell_p}(n) = \left\{ \begin{pmatrix} x_1 \\ \vdots \\ x_n \end{pmatrix} \in \mathbb{R}^+(n), \ |x_1|^p + \cdots |x_n|^p \leqslant 1 \right\} \tag{3.28}$$

$$\mathbb{R}^-_{\ell_q}(n) = \left\{ (x_1, \ldots, x_n) \in \mathbb{R}^-(n), \ |x_1|^q + \cdots |x_n|^q \leqslant 1 \right\}$$

and similarly for $\mathbb{C}^+_{\ell_p}$ and $\mathbb{C}^-_{\ell_q}$.

These are (compact "valuation" cf. Chapter 11) sub-bios of $\mathbb{R}$ (respectively, $\mathbb{C}$). For $p = 2$, they have an involution, and are the "integers" of $\mathbb{R}$ (respectively, $\mathbb{C}$):

The bio $\mathbb{Z}_\mathbb{R} := (\mathbb{R}^-_{\ell_2}, \mathbb{R}^+_{\ell_2}) \subseteq \mathbb{R}$ is the "Real integers", analogue at the "Real prime" of the p-adic integers $\mathbb{Z}_p \subseteq \mathbb{Q}_p$, and is given by the unit real ℓ_2-balls

$$\mathbb{Z}^\pm_\mathbb{R}(n) := \left\{ (x_1, \ldots, x_n) \in \mathbb{R}^n, |x_1|^2 + \cdots |x_n|^2 \leqslant 1 \right\}$$

Similarly, we have the "complex integers" $\mathbb{Z}_\mathbb{C} \subseteq \mathbb{C}$ given by the unit ℓ_2 complex balls.

Note that $\mathbb{Z}^{\mathcal{B}io}_\mathbb{R} = U\mathbb{Z}_\mathbb{R}$, but for the free prop generated by a bio, we have $\mathcal{F}\mathbb{Z}^{\mathcal{B}io}_\mathbb{R} \neq \mathbb{Z}^{\mathcal{P}rop}_\mathbb{R}$, and similarly for $\mathbb{Z}_\mathbb{C}$, so some care is needed!

Contracting the inside of the unit ℓ_2-balls to a point we get quotient bios

$$\begin{aligned} \mathbb{Z}_\mathbb{R} &\twoheadrightarrow \mathbb{F}_\mathbb{R}, \ \mathbb{F}^\pm_\mathbb{R}(n) \equiv S^{n-1} \sqcup \{0\} \\ \mathbb{Z}_\mathbb{C} &\twoheadrightarrow \mathbb{F}_\mathbb{C}, \ \mathbb{F}^\pm_\mathbb{C}(n) \equiv S^{2n-1} \sqcup \{0\} \end{aligned} \tag{3.29}$$

We now have a complete picture of the arithmetical bios, it is obtained by applying the functor U to the same diagram for props (0.18),

$$
\begin{array}{ccccccccc}
 & & & & \mathbb{C} & \supseteq & \mathbb{Z}_{\mathbb{C}} & \twoheadrightarrow & \mathbb{F}_{\mathbb{C}} \\
 & & & & \rotatebox{90}{$\subseteq$} & & \rotatebox{90}{$\subseteq$} & & \rotatebox{90}{$\subseteq$} \\
 & & & & \mathbb{R} & \supseteq & \mathbb{Z}_{\mathbb{R}} & \twoheadrightarrow & \mathbb{F}_{\mathbb{R}} \\
 & & & & \rotatebox{90}{$\subseteq$} & & & & \\
 & & \mathbb{Z} & \subseteq & \mathbb{Q} & & & & \\
 & & \rotatebox{-90}{$\subseteq$} & & \rotatebox{-90}{$\subseteq$} & & & & \\
\mathbb{F}_p & \twoheadleftarrow & \mathbb{Z}_p & \subseteq & \mathbb{Q}_p & & & &
\end{array}
\tag{3.30}
$$

Chapter 4

Commutativity for Bios

Intuitively, the bio $\mathcal{P} = (\mathcal{P}^-, \mathcal{P}^+)$ is **commutative** if operations and cooperations **interchange**: for $q \in \mathcal{P}^+(m)$, $p \in \mathcal{P}^-(n)$,

$$q \circ p \equiv \underbrace{(p, \ldots, p)}_{m} \circ \sigma_{m,n} \circ \underbrace{(q, \ldots, q)}_{n} \tag{4.1}$$

Pictorially, (0.25) is identifier with (0.26) or e.g. $m = 3$, $n = 2$,

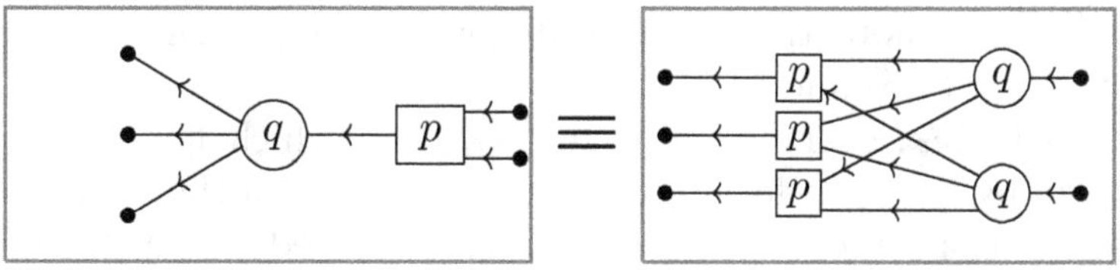

(4.2)

Note though that the elements in (4.1) are no longer elements of $\mathcal{P}$ if $m, n > 1$; they have bi-degree (m, n), and they belong to $(\mathcal{F}\mathcal{P})_{m,n}$. Indeed, formula (4.1) is just the special case of total-commutativity (1.16) with $q_1 = p_0 = 1$. which we called "full commutativity". To get elements of $\mathcal{P}$ we need to "close" (4.1) either on the left or on the right. The simplest way of closing it on the left (respectively, right) is to multiply it on the left (respectively, right) by an element $b \in \mathcal{P}^-(m)$ (respectively, $d \in \mathcal{P}^+(n)$), obtaining

$$b \overleftarrow{\circ} q \circ p = b \circ \underbrace{(p, \ldots, p)}_{m} \circ \sigma_{m,n} \circ \underbrace{(q, \ldots, q)}_{n} \tag{4.3}$$

respectively,

$$q \circ p \overrightarrow{\circ} d = \underbrace{(p, \ldots, p)}_{m} \circ \sigma_{m,n} \circ \underbrace{(q, \ldots, q)}_{n} \circ d \tag{4.4}$$

Note that when $\mathcal{P} = UA$ for a prop A, the identity (4.3) (respectively, (4.4)) follows from the commutativity of A, identity (1.15), by taking $n = 1$ (respectively, $m = 1$) there.

This is all the commutativity we will need, and we let $\mathcal{CBio} \subseteq \mathcal{Bio}$ denote the full subcategory of commutative bios satisfying (4.3) and (4.4).

On the other hand, identity (4.1) is more general than its special cases (4.3), (4.4), and it is quite useful having this more general form. Thinking about the (inductive process of making) formulas in the language of bio, we see that to get this more general commutativity we only need to multiply (4.1) on the left (respectively, right) by an element $b \in \mathcal{P}^-(m')$, $m' \geqslant m$ (respectively, $d \in \mathcal{P}^+(n')$, $n' \geqslant n$), obtaining for $j \geqslant m' - m$ (respectively, $n' - n$),

$$b \overleftarrow{\circ_j} q \circ p = b \overleftarrow{\circ_j} (\underbrace{p, \ldots, p}_{m}) \circ \sigma_{m,n} \circ (\underbrace{q, \ldots, q}_{n}) \tag{4.5}$$

$$q \circ p \overrightarrow{\circ_j} d = (\underbrace{p, \ldots, p}_{m}) \circ \sigma_{m,n} \circ (\underbrace{q, \ldots, q}_{n}) \overrightarrow{\circ_j} d \tag{4.6}$$

We let $\mathcal{C}_f\mathcal{Bio}$ denote the full subcategory of **fully-commutative** Bios satisfying (4.5) and (4.6).

For $\mathcal{P} \in \mathcal{C}_f\mathcal{Bio}$, every element in $(\mathcal{F}_f\mathcal{P})_{n,m}$, which for a general bio is given by the prop-formulas one can make out of the operations and co-operations of $\mathcal{P}$, and is quite complicated, reduces in the fully-commutative case to an expression of the form

$$\begin{aligned} &\left(\bigoplus_{i=1}^{n} p_i\right) \circ \sigma \circ \left(\bigoplus_{j=1}^{m} q_j\right) \\ &p_i \in \mathcal{P}^-(N_i), \quad q_j \in \mathcal{P}^+(M_j), \quad \sigma \in S_M \\ &N_1 + \cdots N_n = M_1 + \cdots + M_m = M \end{aligned} \tag{4.7}$$

Indeed, using (4.5) and (4.6), we can move all the operations p_i (respectively, co-operations q_j) to the left (respectively, right).

For a commutative bio $\mathcal{P}$ the underlying (associative, unital) monoid $\mathcal{P}(1)$ is commutative. Moreover, $\mathcal{P}(1)$ acts **centrally** on $\mathcal{P}^\pm(n)$: for $a \in \mathcal{P}(1)$

$$\begin{aligned} a \cdot p &:= a \circ p = p \circ (a, \ldots, a), \quad p \in \mathcal{P}^-(n) \\ a \cdot q &:= q \circ a = (a, \ldots, a) \circ q, \quad q \in \mathcal{P}^+(n). \end{aligned} \tag{4.8}$$

For a **multiplicative subset** $S \subseteq \mathcal{P}(1)$

$$s_1, s_2 \in S \Longrightarrow s_1 \circ s_2 \in S, \quad \mathbf{1} \in S, \tag{4.9}$$

we have the **localization** of $\mathcal{P}$ at S, $S^{-1}\mathcal{P}$:

$$\begin{aligned} S^{-1}\mathcal{P} &\equiv \mathcal{P}[S^{-1}] \equiv (S^{-1}\mathcal{P}^{-}, S^{-1}\mathcal{P}^{+}) \\ S^{-1}\mathcal{P}^{\pm} &= \mathcal{P}^{\pm} \times S/\sim \end{aligned} \tag{4.10}$$

with $(p'', s'') \sim (p', s')$ iff $s \circ s' \cdot p'' = s \circ s'' \cdot p'$ for some $s \in S$;
one writes $p/s = \frac{1}{s} \circ p$ for the equivalence class $(p, s)/\sim$.

Definition 4.1. We denote by $\mathcal{C}_T\mathcal{B}io \subseteq \mathcal{C}_f\mathcal{B}io$ the full sub-category consisting of the "totally-commutative" bios $\mathcal{P} = (\mathcal{P}^{-}, \mathcal{P}^{+})$, which, in addition to full-commutativity, satisfy the Boardman–Vogt interchange [**BV73**]:

For $p \in \mathcal{P}^{-}(n)$, $p' \in \mathcal{P}^{-}(m)$,

$$p \circ (\underbrace{p', \ldots, p'}_{n}) = p' \circ (\underbrace{p, \ldots, p}_{m}) \circ \sigma_{m,n} \tag{4.11}$$

and symmetrically, for $q \in \mathcal{P}^{+}(n)$, $q' \in \mathcal{P}^{+}(m)$,

$$(\underbrace{q', \ldots, q'}_{n}) \circ q = \sigma_{n,m} \circ (\underbrace{q, \ldots, q}_{m}) \circ q' \tag{4.12}$$

We have the full subcategories

$$\mathcal{CR}ing \subseteq \mathcal{CR}ig \subseteq \mathcal{C}_T\mathcal{B}io \subseteq \mathcal{C}_f\mathcal{B}io \subseteq \mathcal{CB}io \subseteq \mathcal{B}io \tag{4.13}$$

They are all complete and co-complete. All limits are created in $\mathcal{S}et$. Also directed co-limits are created in $\mathcal{S}et$. These categories also have all co-limits, and in particular we have push-outs for a diagram

$$\begin{array}{ccc} B & \longrightarrow & A_0 \\ \big\downarrow & & \\ A_1 & & \end{array} \tag{4.14}$$

But the push-out co-limit will be different taken in the different categories of (4.13).

We will denote these push out by

$$\begin{array}{l}
\text{For } \mathcal{CRing} \text{ and } \mathcal{CRig}\text{:} \quad A_0 \underset{B}{\otimes} A_1 \\
\text{For } \mathcal{C}_T\mathcal{Bio}\text{:} \quad A_0 \underset{B}{\boxtimes}^T A_1 \\
\text{For } \mathcal{C}_f\mathcal{Bio}\text{:} \quad A_0 \underset{B}{\boxtimes}^f A_1 \\
\text{For } \mathcal{CBio}\text{:} \quad A_0 \underset{B}{\boxtimes} A_1 \\
\text{For } \mathcal{Bio}\text{:} \quad A_0 \underset{B}{\amalg} A_1
\end{array} \tag{4.15}$$

We have

$$\begin{aligned}
A_0 \boxtimes_B A_1 &= \left(A_0 \underset{B}{\amalg} A_1\right)^C \\
A_0 \boxtimes_B^f A_1 &= \left(A_0 \boxtimes_B A_1\right)^f \\
A_0 \underset{B}{\boxtimes}^T A_1 &= \left(A_0 \underset{B}{\boxtimes}^f A_1\right)^T
\end{aligned} \tag{4.16}$$

where

$$A \twoheadrightarrow A^C, \text{respectively } A \twoheadrightarrow A^f, \quad A \twoheadrightarrow A^T \tag{4.17}$$

is the maximal (respectively, fully/totally) commutative quotient of A giving the left adjoint of the embedding $\mathcal{CBio} \subseteq \mathcal{Bio}$, respectively, $\mathcal{C}_f\mathcal{Bio} \subseteq \mathcal{CBio}$, $\mathcal{C}_T\mathcal{Bio} \subseteq \mathcal{C}_f\mathcal{Bio}$.

We have the commutative diagram of adjunctions where $\mathcal{F}$ (respectively, $\mathcal{F}_T, \mathcal{F}_f, \mathcal{F}_C$) is the free (respectively, totally-, full-,commutative) prop generated by a bio.

$$\begin{array}{ccccccc}
P^T \longleftarrow\!\shortmid P & , & P^f \longleftarrow\!\shortmid P & , & P^C \longleftarrow\!\shortmid P & & \\
\mathcal{C}_T\mathcal{Prop} & \leftrightarrows & \mathcal{C}_f\mathcal{Prop} & \leftrightarrows & \mathcal{CProp} & \leftrightarrows & \mathcal{Prop} \\
\mathcal{F}_T \uparrow\downarrow U & & \mathcal{F}_f \uparrow\downarrow U & & \mathcal{F}_C \uparrow\downarrow U & & \mathcal{F} \uparrow\downarrow U \\
\mathcal{C}_T\mathcal{Bio} & \leftrightarrows & \mathcal{C}_f\mathcal{Bio} & \leftrightarrows & \mathcal{CBio} & \leftrightarrows & \mathcal{Bio}
\end{array} \tag{4.18}$$

Example 4.1. Taking $B = \mathbb{F}$, $A_i = \mathbb{Z}$, in (4.14) we get

$$\mathbb{Z} \coprod_{\mathbb{F}} \mathbb{Z} \twoheadrightarrow \mathbb{Z} \underset{\mathbb{F}}{\boxtimes} \mathbb{Z} \twoheadrightarrow \mathbb{Z} \underset{\mathbb{F}}{\boxtimes}^f \mathbb{Z} \twoheadrightarrow \mathbb{Z} \underset{\mathbb{F}}{\boxtimes}^T \mathbb{Z} \overset{\sim}{\twoheadrightarrow} \mathbb{Z} \tag{4.19}$$

The last map, given by the diagonal homomorphism, is an isomorphism: If we require total-commutativity the arithmetical surface again reduces to its diagonal, and $\mathbb{Z} \boxtimes_{\mathbb{F}}^{T} \mathbb{Z} = \mathbb{Z}$. Indeed, $\mathbb{Z}$ is generated over $\mathbb{F}$ by $v = (1,1)$, $v^t = \begin{pmatrix} 1 \\ 1 \end{pmatrix}$, (and (-1)), and total commutativity implies

$$\left.\begin{array}{l} (v \boxtimes 1) \circ (1 \boxtimes v, 1 \boxtimes v) \circ \big((1,0),(0,1)\big) \\ \equiv \\ (1 \boxtimes v) \circ (v \boxtimes 1, v \boxtimes 1) \circ \big((1,0),(0,1)\big) \end{array}\right\} \implies v \boxtimes 1 = 1 \boxtimes v \tag{4.20}$$

pictorially

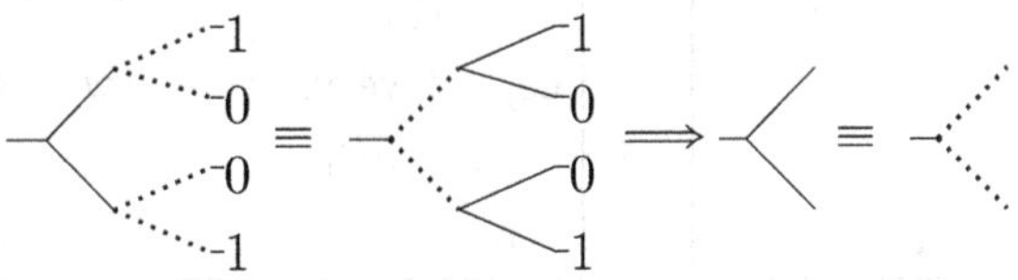

(4.21)

This proof is the proof of Hurwitz's Lemma from algebraic topology: given a pointed set with two binary operations $*$, $\star : X \times X \to X$ that **commute**

$$(x_1 \star x_2) * (x_3 \star x_4) \equiv (x_1 * x_2) \star (x_3 * x_4),$$

with the distinguished point $o \in X$ being an identity for both

$$x \star o = o \star x = x * o = o * x = x,$$

the operations are one and the same, and they are commutative

$$x \star y = x * y = y * x.$$

Thus to have a non-trivial Arithmetical Surface we have to give up total-commutativity. This is the reason we introduce the weaker notions of (full)-commutativity. We have similar phenomena as in (4.19) for props.

The elements of $\mathbb{N} \underset{\mathbb{F}}{\boxtimes}^{f} \mathbb{N} = (\mathbb{N} \underset{\mathbb{F}}{\amalg} \mathbb{N})^f$, respectively

$$\mathbb{Z} \underset{\mathbb{F}[\pm 1]}{\boxtimes}^{f} \mathbb{Z} = (\mathbb{Z} \underset{\mathbb{F}[\pm 1]}{\amalg} \mathbb{Z})^f,$$

the categorical sums taken in the category of **fully-commutative props**, can be represented as data

$$(Y_1, \dots, Y_n; \sigma; X_1, \dots, X_m) \tag{4.22}$$

where $Y_1, \ldots, Y_n, X_1, \ldots, X_m$ are ordered sequences of finite rooted trees, and σ gives a bijection between the union of their leaves:

$$\sigma : \coprod_{j=1}^{m} \partial X_j, \xrightarrow{\sim} \coprod_{i=1}^{n} \partial Y_i \tag{4.23}$$

respectively, and for each pair $x \in \partial X_j$, $y \in \partial Y_i$, $\sigma(x) = y$, σ also assigns a sign $\mathcal{E}(x, y) \in \{+1, -1\}$. Indeed, it is generated by the long vectors (2.25), $v_n^{(0)}, v_n^{(1)}, (v_n^{(0)})^t, (v_n^{(1)})^t$ where the upper script 0/1 indicate that it is coming from the left/right factors of the categorical sums; we can move using full-commutativity all the $v_n^{(0)}, v_n^{(1)}$ (respectively, $(v_n^{(0)})^t, (v_n^{(1)})^t$) to the left (respectively, right). Multiplying the $v_n^{(0)}, v_n^{(1)}$ (respectively, $(v_n^{(0)})^t, (v_n^{(1)})^t$) among themselves, we get the trees Y_i's (respectively, X_j's), where a vertex of height h above the root with n offsprings represents $v_n^{(h \bmod 2)}$ (respectively, $(v_n^{(h \bmod 2)})^t$). We may assume that the only vertices with a unique offspring are the roots (as such a vertex represents the identity $v_1^{(\epsilon)} = (1)$). The operations are given by

$$\begin{aligned}(Y_1, \ldots, Y_n, \sigma; X_1, \ldots, X_m) \oplus (Y_1', \ldots, Y_{n'}', \sigma'; X_1', \ldots, X_{m'}') :=\\ (Y_1, \ldots, Y_n, Y_1', \ldots, Y_{n'}', \sigma \coprod \sigma'; X_1, \ldots, X_m, X_1', \ldots, X_{m'}')\end{aligned} \tag{4.24}$$

$$\begin{aligned}(Y_1, \ldots, Y_n; \sigma; X_1, \ldots, X_m) \circ (Z_1, \ldots, Z_m; \tau; W_1, \ldots, W_\ell) :=\\ \left(Y_1 \underset{\sigma}{\ltimes} (Z_j) \ldots Y_n \underset{\sigma}{\ltimes} (Z_j);\ \sigma \circ \tau; W_1 \underset{\tau}{\ltimes} (X_j) \ldots W_\ell \underset{\tau}{\ltimes} (X_j)\right)\end{aligned}$$

Here $Y_i \underset{\sigma}{\ltimes} (Z_j)$ is the tree Y_i with the trees Z_j glued to its leaves, where at $y \in \partial Y_i$ with $\sigma^{-1}(y) = x \in \partial X_{j_0}$, we glue the tree Z_{j_0} (and similarly $W_k \underset{\tau}{\ltimes} (X_j)$ is the tree W_k with the tree X_{j_0} glued at $w \in \partial W_k$ if $\tau(w) \in \partial Z_{j_0}$).

We say the data in (4.22) have **degree** $\sum_j \#\partial X_j \equiv \sum_i \#\partial Y_i$, and for $Z \in \mathbb{N} \underset{\mathbb{F}}{\boxtimes}^f \mathbb{N}$, respectively, $\mathbb{Z} \underset{\mathbb{F}[\pm 1]}{\boxtimes}^f \mathbb{Z}$, we define $\deg(Z)$ to be the minimal degree of a data representing it. We have

$$\deg(Z \circ Z') \leqslant \deg(Z) \cdot \deg(Z')$$

$$\deg(Z_1 \oplus Z_2) \equiv \deg(Z_1) + \deg(Z_2)$$

The diagonal homomorphism

$$\Delta : \mathbb{N} \underset{\mathbb{F}}{\boxtimes^f} \mathbb{N} \longrightarrow\!\!\!\!\!\rightarrow \mathbb{N}, \quad \mathbb{Z} \underset{\mathbb{F}[\pm 1]}{\boxtimes^f} \mathbb{Z} \longrightarrow\!\!\!\!\!\rightarrow \mathbb{Z},$$

is given by $\Delta(Y_0, \ldots, Y_n, \sigma, X_1, \ldots, X_m) = (a_{ij})$, where a_{ij} is the number of paths going from the root of X_j to the root of Y_i in the graph made from the union of the X_j's and Y_i's glued along σ (respectively, taking the sign of σ into consideration):

$$a_{ij} = \#\{(x, y) \in \partial X_j \times \partial Y_i, \sigma(x) = y\}$$

respectively,

$$\begin{aligned} a_{ij} = \#\{(x, y) \in \partial X_j \times \partial Y_i, \sigma(x) = y, \mathcal{E}(x, y) = +1\} \\ -\#\{(x, y) \in \partial X_j \times \partial Y_i, \sigma(x) = y, \mathcal{E}(x, y) = -1\} \end{aligned}$$

The set of elements of bi-degree $(1, 1)$, $(\mathbb{Z} \underset{\mathbb{F}[\pm 1]}{\boxtimes^f} \mathbb{Z})_{1,1}$, has **two** operations of addition, cf. (2.28), making it into a commutative ring in two different ways:

$$Z_1 +_i Z_2 = v^{(i)} \circ (Z_1 \oplus Z_2) \circ (v^{(i)})^t$$

$$v^{(i)} = (1, 1) \in (\mathbb{Z} \underset{\mathbb{F}[\pm 1]}{\boxtimes^f} \mathbb{Z})_{1,2}$$

where $i = 0/1$ for the left/right elements. Moreover, it has **two** commuting involutions: the one involution $Z \mapsto Z^t$, coming from taking transpose of matrices $(\mathbb{Z} \underset{\mathbb{F}[\pm 1]}{\boxtimes^f} \mathbb{Z})_{n,m} \xrightarrow{\sim} (\mathbb{Z} \underset{\mathbb{F}[\pm 1]}{\boxtimes^f} \mathbb{Z})_{m,n}$, with $n = m = 1$, preserves these ring structures,

$$(Z_1 +_i Z_2)^t = Z_1^t +_i Z_2^t$$

the other involution $Z \mapsto Z^*$, coming from permuting the left and right factors in $\mathbb{Z} \underset{\mathbb{F}[\pm 1]}{\boxtimes^f} \mathbb{Z}$, preserves multiplication, and permutes the two additions

$$(Z_1 +_i Z_2)^* = Z_1^* +_{1-i} Z_2^*$$

Chapter 5

Ideals and Primes

For props and bios A there are different notions of "ideals", beginning with the "equivalence ideal" (subset of $A \prod A$), the "prop or bio ideals" (subsets of A), and for a bio the "ideals" subset of $A(1)$ (or of $P_{1,1}$ if $A = UP$ for a prop $P \in \mathcal{Prop}$).

Definition 5.1. For a prop (respectively, bio) A, an **equivalence-ideal** is an equivalence relation $R \subseteq A \prod A$, that is also a sub prop (respectively, bio).

We write $a_1 \underset{R}{\sim} a_2$, or just $a_1 \sim a_2$ if R is clear, for $(a_1, a_2) \in R$.

We write A/R or $A/_{\sim}$ for the collection of equivalence classes. We have

$$a_1 \sim a_1', \quad a_2 \sim a_2' \implies a_1 \circ a_2 \sim a_1' \circ a_2' \tag{5.1}$$

and for props also

$$a_1 \sim a_1', \quad a_2 \sim a_2' \implies a_1 \oplus a_2 \sim a_1' \oplus a_2' \tag{5.2}$$

On A/R there is a unique structure of a prop (respectively, bio) so that the canonical map is a homomorphism

$$\pi_R : A \twoheadrightarrow A/_R \tag{5.3}$$

Example 5.1. For a map $\varphi \in \mathcal{Prop}(A, B)$, (respectively, $\mathcal{Bio}(A, B)$) its kernel is an equivalence ideal

$$\ker(\varphi) = A \prod_{B} A = \left\{(a, a') \in A \prod A,\ \varphi(a) = \varphi(a')\right\} \tag{5.4}$$

The image of φ is the sub-prop (respectively, bio)

$$\varphi(A) \subseteq B \tag{5.5}$$

and we have the factorization of φ

$$\begin{array}{ccc} A & \xrightarrow{\quad \varphi \quad} & B \\ \Big\downarrow\!\!\downarrow & & \Big\uparrow \cup \\ A/_{\ker(\varphi)} & \xrightarrow{\quad \sim \quad} & \varphi(A) \end{array} \tag{5.6}$$

Remark 5.1. Before we define the "ideals" of bios that are of interest to us, we mention that for a prop A there is a natural notion, we will call it a **prop-ideal**. A prop-ideal of A is a collection of subsets $\mathfrak{a} = \{\mathfrak{a}_{n,m}\}$, $\mathfrak{a}_{n,m} \subseteq A_{n,m}$, such that

$$\begin{aligned} &\text{(i)}\ \ A_{n',n} \circ \mathfrak{a}_{n,m} \circ A_{m,m'} \subseteq \mathfrak{a}_{n',m'} \\ &\text{(ii)}\ \mathfrak{a}_{n_1,m_1} \oplus \mathfrak{a}_{n_2,m_2} \subseteq \mathfrak{a}_{n_1+n_2,m_1+m_2} \end{aligned} \tag{5.7}$$

Similarly, for a bio P there is the notion of a **bio-ideal**, it is a collection of subsets $\mathfrak{a} = \{\mathfrak{a}^{\pm}(n)\}$, $\mathfrak{a}^{\pm}(n) \subseteq P^{\pm}(n)$, that is closed under multiplications and actions on both the left and right sides:

$$\begin{aligned} &\circ:\quad P^-(n) \times \mathfrak{a}^-(k_1) \times \cdots \times \mathfrak{a}^-(k_n) \longrightarrow \mathfrak{a}^-(k_1 + \cdots + k_n) \\ &\circ:\quad \mathfrak{a}^-(n) \times P^-(k_1) \times \cdots \times P^-(k_n) \longrightarrow \mathfrak{a}^-(k_1 + \cdots + k_n) \\ &\mathfrak{a}^+(k_1 + \cdots + k_n) \longleftarrow \mathfrak{a}^+(k_1) \times \cdots \times \mathfrak{a}^+(k_n) \times P^+(n)\ :\circ \\ &\mathfrak{a}^+(k_1 + \cdots + k_n) \longleftarrow P^+(k_1) \times \cdots \times P^+(k_n) \times \mathfrak{a}^+(n)\ :\circ \\ &\overleftarrow{\circ}:\quad P^-(k_1 + \cdots + k_n) \times \mathfrak{a}^+(k_1) \times \cdots \times \mathfrak{a}^+(k_n) \longrightarrow \mathfrak{a}^-(n) \\ &\overleftarrow{\circ}:\quad \mathfrak{a}^-(k_1 + \cdots + k_n) \times P^+(k_1) \times \cdots \times P^+(k_n) \longrightarrow \mathfrak{a}^-(n) \\ &\mathfrak{a}^+(n) \longleftarrow \mathfrak{a}^-(k_1) \times \cdots \times \mathfrak{a}^-(k_n) \times P^+(k_1 + \cdots + k_n) : \overrightarrow{\circ} \\ &\mathfrak{a}^+(n) \longleftarrow P^-(k_1) \times \cdots \times P^-(k_n) \times \mathfrak{a}^+(k_1 + \cdots + k_n)\ : \overrightarrow{\circ} \end{aligned} \tag{5.8}$$

Definition 5.2. For a bio A, (and hence for a prop P taking $A = UP$), an **ideal** is a subset $\mathfrak{a} \subseteq A(1)$, that is **closed under linear combinations:**

$$\text{For } a_1, \ldots, a_n \in \mathfrak{a}\,,\ b \in A^-(n)\,,\ d \in A^+(n) \text{ we have } b \circ (a_i) \circ d \in \mathfrak{a} \tag{5.9}$$

Note that if $1 \in \mathfrak{a}$, then $\mathfrak{a} \equiv A(1)$ is the trivial ideal, and we usually mean by an "ideal" a non-trivial ideal. The intersection of ideals $\mathfrak{a}_i \subseteq A(1)$ is again an ideal $\bigcap_i \mathfrak{a}_i \subseteq A(1)$. Given a subset $\mathbb{a} \subseteq A(1)$, the intersection of all the ideals containing $\mathbb{a}$ is the **ideal generated by** $\mathbb{a}$

$$\langle \mathbb{a} \rangle = \bigcap_{\mathbb{a} \subseteq \mathfrak{a} \subseteq A(1) \text{ ideal}} \mathfrak{a} \tag{5.10}$$

Since a linear combination of linear combinations is again a linear combination, we can describe $\langle \mathbb{a} \rangle$ also as the collection of all linear combinations

$$\langle \mathbb{a} \rangle = \bigcup_n \{ b \circ (a_i) \circ d, b \in A^-(n), d \in A^+(n), a_i \in \mathbb{a} \} \tag{5.11}$$

Given an equivalence-ideal $R \subseteq A \Pi A$, we get the ideal:

$$V(R) := R_{1,1} \cap (A_{1,1} \times \{o_{1,1}\}) = \{ a \in A(1),\ a \underset{R}{\sim} o \} \tag{5.12}$$

Given an ideal $\mathfrak{a} \subseteq A(1)$, we get the equivalence-ideal generated by it

$$E(\mathfrak{a}) := \bigcap_{\mathfrak{a} \times \{o_{1,1}\} \subseteq R \subseteq A \Pi A \text{ equiv ideal}} R \tag{5.13}$$

This gives a Galois correspondence between ideals and equivalence-ideals:

$$\begin{array}{c} R \in \mathcal{E}quiv(A) \\ E \left(\uparrow \quad \downarrow \right) V \\ \mathfrak{a} \in \mathcal{I}deal(A) \end{array} \tag{5.14}$$

$$\mathfrak{a} \subseteq V(R) \iff E(\mathfrak{a}) \subseteq R \tag{5.15}$$

$$\mathfrak{a} \subseteq VE(\mathfrak{a}) \quad EV(R) \subseteq R \tag{5.16}$$

Remark 5.2. For a prop A, this adjunction factorize through the set of prop ideals of A

$$\mathcal{Ideal}(A) \underset{<>_{1,1}}{\overset{F}{\rightleftarrows}} \text{prop-Id}(A) \underset{V}{\overset{E}{\rightleftarrows}} \mathcal{Equiv}(A) \tag{5.17}$$

Similarly for a bio A this adjunction factorize through the set of bio-ideals of A.

Given ideals $\mathfrak{a}_1, \mathfrak{a}_2$ (respectively, $\mathfrak{a}_i, i \in I$), we have their sum

$$\mathfrak{a}_1 + \mathfrak{a}_2 := \langle \mathfrak{a}_1 \cup \mathfrak{a}_2 \rangle \tag{5.18}$$

$$\sum_{i \in I} \mathfrak{a}_i := \left\langle \bigcup_{i \in I} \mathfrak{a}_i \right\rangle \tag{5.19}$$

We have also the **product** of ideals

$$\mathfrak{a}_1 \cdot \mathfrak{a}_2 := \langle \{a_1 \circ a_2, a_i \in \mathfrak{a}_i\} \rangle \tag{5.20}$$

These satisfy the usual properties, forming a commutative rig for A central

$$\mathfrak{a}_1 + (\mathfrak{a}_2 + \mathfrak{a}_2) = \mathfrak{a}_1 + \mathfrak{a}_2 + \mathfrak{a}_2 = (\mathfrak{a}_1 + \mathfrak{a}_2) + \mathfrak{a}_2$$
$$\mathfrak{a}_1 + \mathfrak{a}_2 = \mathfrak{a}_2 + \mathfrak{a}_1, \quad \mathfrak{a} + \{o\} = \mathfrak{a}$$
$$\mathfrak{a}_1 \cdot (\mathfrak{a}_2 \cdot \mathfrak{a}_2) = \mathfrak{a}_1 \cdot \mathfrak{a}_2 \cdot \mathfrak{a}_3 = (\mathfrak{a}_1 \cdot \mathfrak{a}_2) \cdot \mathfrak{a}_2$$
$$\mathfrak{a}_1 \cdot \mathfrak{a}_2 = \mathfrak{a}_2 \cdot \mathfrak{a}_1, \quad \mathfrak{a} \cdot \langle 1 \rangle = \mathfrak{a}$$
$$\mathfrak{a} \cdot (\mathfrak{b}_1 + \mathfrak{b}_2) = (\mathfrak{a} \cdot \mathfrak{b}_1) + (\mathfrak{a} \cdot \mathfrak{b}_2), \quad (\mathfrak{a}_1 + \mathfrak{a}_2) \cdot \mathfrak{b} = (\mathfrak{a}_1 \cdot \mathfrak{b}) + (\mathfrak{a}_2 \cdot \mathfrak{b})$$
$$\mathfrak{a} \cdot \{0\} = \{0\} = \{0\} \cdot \mathfrak{a}$$

The union of a chain of (non-trivial) ideals is again a (non-trivial) ideal.

An application of Zorn's lemma gives therefore,

Lemma 5.1. *There exists a maximal non-trivial ideal.*

We write $\mathcal{max}(A)$ for the collection of maximal ideals of A.

Definition 5.3. An ideal $\mathfrak{p} \subseteq A(1)$ will be called **prime** if $S_{\mathfrak{p}} := A(1) \backslash \mathfrak{p}$ is multiplicative:

$$s_1, s_2 \in S_{\mathfrak{p}} \Longrightarrow s_1 \circ s_2 \in S_{\mathfrak{p}}; \quad 1 \in S_{\mathfrak{p}}$$

We write $\mathcal{spec}(A)$ for the collection of primes.

Lemma 5.2. *Let A be a commutative bio. We have*

$$\phi \neq \mathcal{max}(A) \subseteq \mathcal{spec}(A)$$

i.e. every maximal ideal is prime.

Proof. Let $\mathfrak{m} \subseteq A(1)$ be maximal and let $s_1, s_2 \in A(1) \backslash \mathfrak{m}$. The ideal generated by $\mathfrak{m}$ and s_1, properly contains $\mathfrak{m}$ and so must be trivial, so we have a linear combination

$$1 = b_1 \circ (a_i^{(1)}) \circ d_1, \quad \text{with } a_i^{(1)} \in \mathfrak{m} \text{ or } a_i^{(1)} = s_1, \quad i = 1, \ldots, n_1$$

Similarly,

$$1 = b_2 \circ (a_j^{(2)}) \circ d_2, \quad \text{with } a_j^{(2)} \in \mathfrak{m} \text{ or } a_j^{(2)} = s_2, \quad j = 1, \ldots, n_2$$

We obtain

$$\begin{aligned} 1 = 1 \circ 1 &= b_1 \circ (a_i^{(1)}) \circ d_1 \circ b_2 \circ (a_j^{(2)}) \circ d_2 \quad \text{and by commutativity} \\ &= b_1 \circ (a_i^{(1)}) \circ \underbrace{(b_2, \ldots, b_2)}_{n_1} \circ \sigma_{n_1, n_2} \circ \underbrace{(d_1, \ldots, d_1)}_{n_2} \circ (a_j^{(2)}) \circ d_2 \\ &= (b_1 \circ (b_2, \ldots, b_2)) \circ (a_i^{(1)} \cdot a_j^{(2)}) \circ (\sigma_{n_1, n_2} \circ (d_1, \ldots, d_1) \circ d_2) \end{aligned}$$

is a linear combination of the $a_i^{(1)} \cdot a_j^{(2)}$. But either $a_i^{(1)} \cdot a_j^{(2)} \in \mathfrak{m}$ or $a_i^{(1)} \cdot a_j^{(2)} = s_1 \cdot s_2$, and since $1 \notin \mathfrak{m}$ we must have $s_1 \cdot s_2 \notin \mathfrak{m}$. □

Given a map $\varphi \in \mathcal{Bio}(A,B)$, (respectively, $\mathcal{Prop}(A,B)$) the pull back of an ideal is an ideal, and the pull back of a prime is a prime,

$$\begin{array}{ccc} \mathcal{Ideal}(A) & \longleftarrow & \mathcal{Ideal}(B) : \varphi^* \\ \text{IU} & & \text{IU} \\ \mathit{spec}(A) & \longleftarrow & \mathit{spec}(B) : \varphi^* \end{array} \tag{5.21}$$

$$\varphi_1^{-1}(\mathfrak{b}) = \varphi^*(\mathfrak{b}) \;\leftarrow\!\!\!\sim\!\sim\!\sim\!\!\!\dashv\; \mathfrak{b} \subseteq B(1)$$

For an ideal $\mathfrak{a} \subseteq A(1)$, we let $\varphi_*(\mathfrak{a}) = \langle \varphi(\mathfrak{a}) \rangle$ the ideal generated by its image. We get Galois adjunction

$$\mathcal{Ideal}(A) \underset{\varphi^*}{\overset{\varphi_*}{\rightleftarrows}} \mathcal{Ideal}(B)$$

$$\mathfrak{a} \subseteq \varphi^*(\mathfrak{b}) \iff \varphi_*(\mathfrak{a}) \subseteq \mathfrak{b}$$
$$\mathfrak{a} \subseteq \varphi^*\varphi_*(\mathfrak{a}), \qquad \varphi_*\varphi^*(\mathfrak{b}) \subseteq \mathfrak{b}$$

and induced bijection

$$\begin{array}{ccc} \{\mathfrak{a} \in \mathcal{Ideal}(A), \mathfrak{a} = \varphi^*\varphi_*(\mathfrak{a})\} & & \{\mathfrak{b} \in \mathcal{Ideal}(B), \mathfrak{b} = \varphi_*\varphi^*(\mathfrak{b})\} \\ ||| & \overset{\sim}{\longleftrightarrow} & ||| \\ \{\varphi^*(\mathfrak{b}), \mathfrak{b} \in \mathcal{Ideal}(B)\} & & \{\varphi_*(\mathfrak{a}), \mathfrak{a} \in \mathcal{Ideal}(A)\} \end{array}$$

The map φ_* is a homomorphism of rigs

$$\varphi_*(\mathfrak{a}_1 + \mathfrak{a}_2) = \varphi_*(\mathfrak{a}_1) + \varphi_*(\mathfrak{a}_2), \quad \varphi_*(\mathfrak{a}_1 \cdot \mathfrak{a}_2) = \varphi_*(\mathfrak{a}_1) \cdot \varphi_*(\mathfrak{a}_2)$$
$$\varphi_*(0) = 0, \qquad \varphi_*(\langle 1 \rangle) = \langle 1 \rangle$$

For an ideal $\mathfrak{a} \subseteq A(1)$, and the equivalence ideal $E(\mathfrak{a})$ generated by $\mathfrak{a} \times \{0\}$, we write $A/\mathfrak{a} := A/_{E(\mathfrak{a})}$, and let $\pi : A \twoheadrightarrow A/\mathfrak{a}$ the canonical homomorphism. We have $\pi_*\pi^* = \mathrm{id}_{\mathcal{Ideal}(A/_\mathfrak{a})}$, and the image of π^* are the ideals $\mathfrak{b} \subseteq A(1)$, $\mathfrak{b} = \pi^*\pi_*(\mathfrak{b})$, $\mathfrak{b}$ is "$E(\mathfrak{a})$-stable", and we get

$$\pi^* : \mathcal{Ideal}(A/\mathfrak{a}) \xrightarrow{\sim} \{\mathfrak{b} \in \mathcal{Ideal}(A), E(\mathfrak{a})\text{-stable}\}$$

The following lemma is a generalization of Lemma 5.2.

Lemma 5.3. *For a commutative bio A, an ideal $\mathfrak{a} \subseteq A(1)$, and an element $f \in A(1)$, such that $f^m \notin \mathfrak{a}$ for all $m \in \mathbb{N}$, a maximal element of the set*

$$\{\mathfrak{b} \in \mathcal{Ideal}(A), \mathfrak{b} \supseteq \mathfrak{a}, f^m \notin \mathfrak{b} \text{ for all } m \in \mathbb{N}\} \tag{5.22}$$

is prime.

Proof. Let $\mathfrak{p}$ be a maximal element of the set (5.22), and let $s_1, s_2 \in A(1)\backslash\mathfrak{p}$. The ideal generated by $\mathfrak{p}$ and s_i properly contains $\mathfrak{p}$, and by maximally of $\mathfrak{p}$ we must have $f^{m_i} \in \langle \mathfrak{p}, s_i \rangle$. Thus we have some linear combinations

$$f^{m_1} = b_1 \circ (a_i^{(1)}) \circ d_1 \text{ , with } a_i^{(1)} \in \mathfrak{p} \text{ or } a_i^{(1)} = s_1 \text{ , } i = 1, \ldots, n_1$$
$$f^{m_2} = b_2 \circ (a_j^{(2)}) \circ d_2 \text{ , with } a_j^{(2)} \in \mathfrak{p} \text{ or } a_j^{(2)} = s_2 \text{ , } j = 1, \ldots, n_2$$

We obtain

$$f^{m_1+m_2} = f^{m_1} \circ f^{m_2} = b_1 \circ (a_i^{(1)}) \circ d_1 \circ b_2 \circ (a_j^{(2)}) \circ d_2$$

and by commutativity:

$$= b_1 \circ (a_i^{(1)}) \circ (\underbrace{b_2, \ldots, b_2}_{n_1}) \circ \sigma_{n_1 n_2} \circ (\underbrace{d_1, \ldots, d_1}_{n_2}) \circ (a_j^{(2)}) \circ d_2$$
$$= (b_1 \circ (b_2, \ldots, b_2)) \circ (a_i^{(1)} \circ a_j^{(2)}) \circ (\sigma_{n_1, n_2}(d_1, \ldots, d_1) \circ d_2)$$

a linear combination of $a_i^{(1)} \circ a_j^{(2)}$, but $a_i^{(1)} \circ a_j^{(2)} \in \mathfrak{p}$ or

$$a_i^{(1)} \circ a_j^{(2)} = s_1 \circ s_2,$$

and since $f^{m_1+m_2} \notin \mathfrak{p}$ we must have $s_1 \circ s_2 \notin \mathfrak{p}$. □

Note that the set (5.22) is not empty (it contains $\mathfrak{a}$), and for a chain of ideals in it the union will be in it, so by Zorn's lemma there always will be such a maximal element, and so we will always find a prime $\mathfrak{p}$ with

$$\mathfrak{p} \supseteq \mathfrak{a}, \text{ and } f \notin \mathfrak{p}$$

Lemma 5.4. *For $A \in \mathcal{CBio}$, $\mathfrak{a} \in \mathcal{Ideal}(A)$, the* **radical**

$$\sqrt{\mathfrak{a}} := \{f \in A(1), f^m \in \mathfrak{a} \text{ for } m \gg 1\} \equiv \bigcap_{\mathfrak{a} \subseteq \mathfrak{p} \text{ prime}} \mathfrak{p}$$

is equal to the intersections of the prime ideals containing $\mathfrak{a}$.

Proof. If $f \in \sqrt{\mathfrak{a}}$, say $f^m \in \mathfrak{a}$, then for any prime $\mathfrak{p} \supseteq \mathfrak{a}$, $f^m \in \mathfrak{p}$, so $f \in \mathfrak{p}$; thus $\sqrt{\mathfrak{a}} \subseteq \bigcap_{\mathfrak{a}\subseteq\mathfrak{p} \text{ prime}} \mathfrak{p}$. If $f \notin \sqrt{\mathfrak{a}}$, so $f^m \notin \mathfrak{a}$ for all m, then an element $\mathfrak{p}$ maximal in the set (5.22), exist by Zorn's lemma, is a prime $\mathfrak{p} \supseteq \mathfrak{a}$, and $f \notin \mathfrak{p}$, so $f \notin \bigcap_{\mathfrak{a}\subseteq\mathfrak{p} \text{ prime}} \mathfrak{p}$. □

Chapter 6

The Spectrum

Let $A \in \mathcal{CBio}$ be a commutative bio (e.g. $A = UP$, $P \in \mathcal{CProp}$), $\mathit{spec}(A)$ the set of prime ideals. For a subset $\mathfrak{a} \subseteq A(1)$ put

$$V(\mathfrak{a}) := \{\mathfrak{p} \in \mathit{spec}(A), \mathfrak{a} \subseteq \mathfrak{p})\} \tag{6.1}$$

We have $V(\mathfrak{a}) = V(\langle\mathfrak{a}\rangle)$ and we may assume $\mathfrak{a}$ is an ideal.

We have

$$V(1) = \phi, \quad V(\{0\}) = \mathit{spec}(A) \tag{6.2}$$

$$V\left(\sum_{i \in I} \mathfrak{a}_i\right) = \bigcap_{i \in I} V(\mathfrak{a}_i) \tag{6.3}$$

$$V(\mathfrak{a}_1 \cdot \mathfrak{a}_2) = V(\mathfrak{a}_1) \cup V(\mathfrak{a}_2) \tag{6.4}$$

Thus we have the **Zariski topology** on $\mathit{spec}(A)$, with closed sets the $V(\mathfrak{a})$, $\mathfrak{a} \in \mathcal{Ideal}(A)$.

We also have a basis for the topology by **basic open sets**, for $f \in A(1)$,

$$D_A(f) := \mathit{spec}(A) \backslash V(f) = \{\mathfrak{p} \in \mathit{spec}(A), f \notin \mathfrak{p}\} \tag{6.5}$$

$$D_A(f_1) \cap D_A(f_2) = D_A(f_1 \circ f_2) \tag{6.6}$$

$$\mathit{spec}(A) \backslash V(\mathfrak{a}) = \bigcup_{f \in \mathfrak{a}} D_A(f) \tag{6.7}$$

For a subset $X \subseteq \mathit{spec}(A)$ we have the associated ideal

$$I(X) := \bigcap_{\mathfrak{p} \in X} \mathfrak{p} \tag{6.8}$$

We obtain the Galois adjunction

$$\{X \subseteq \mathit{spec}(A) \text{ subset}\} \underset{I}{\overset{V}{\leftrightarrows}} \mathit{Ideal}(A) \tag{6.9}$$

$$X_1 \subseteq X_2 \Longrightarrow I(X_1) \supseteq I(X_2) \tag{6.10}$$

$$V(\mathfrak{a}_1) \subseteq V(\mathfrak{a}_2) \Longleftarrow \mathfrak{a}_1 \supseteq \mathfrak{a}_2 \tag{6.11}$$

$$X \subseteq V(\mathfrak{a}) \Longleftrightarrow I(X) \supseteq \mathfrak{a} \tag{6.12}$$

$$X \subseteq VI(X), \qquad IV(\mathfrak{a}) \supseteq \mathfrak{a} \tag{6.13}$$

Note that by Lemma 5.4,

$$IV(\mathfrak{a}) = \bigcap_{\mathfrak{p} \in V(\mathfrak{a})} \mathfrak{p} = \bigcap_{\mathfrak{a} \subseteq \mathfrak{p}} \mathfrak{p} = \sqrt{\mathfrak{a}} \text{ the radical of } \mathfrak{a} \tag{6.14}$$

We also have

$$VI(X) = \overline{X} \text{ the closure of } X \text{ in the Zariski topology} \tag{6.15}$$

Indeed, $X \subseteq VI(X)$, and $VI(X)$ closed, so $\overline{X} \subseteq VI(X)$. On the other hand, for any closed set $V(\mathfrak{a}) \supseteq X$, we have $IV(\mathfrak{a}) \subseteq I(X)$, hence $V(\mathfrak{a}) = VIV(\mathfrak{a}) \supseteq VI(X)$.

Corollary 6.1. *For $A \in \mathcal{CBio}$, we have a bijection between closed subsets of $\mathit{spec}(A)$ and radical ideals*

$$\begin{aligned} \{X \subseteq \mathit{spec}(A), X = \overline{X}\} &\xleftrightarrow{\sim} \{\mathfrak{a} \in \mathit{Ideal}(A), \mathfrak{a} = \sqrt{\mathfrak{a}}\} \\ \{\mathfrak{a} \subseteq \mathfrak{p}\} = V(\mathfrak{a}) &\leftarrow\!\!\sim\!\!\sim\!\!\dashv \mathfrak{a} \\ X &\rightsquigarrow I(X) = \bigcap_{\mathfrak{p} \in X} \mathfrak{p} \end{aligned} \tag{6.16}$$

Note that for a radical ideal $\mathfrak{a} = \sqrt{\mathfrak{a}}$, and for $f \in A(1)$, we have

$$D_A(f) \cap V(\mathfrak{a}) \neq \phi \Longleftrightarrow \exists \mathfrak{p} \supseteq \mathfrak{a}, \ \mathfrak{p} \not\ni f \ \Longleftrightarrow \ f \notin \bigcap_{\mathfrak{a} \in \mathfrak{p}} \mathfrak{p} = \sqrt{\mathfrak{a}} = \mathfrak{a}$$

The close set $V(\mathfrak{a})$ is **irreducible** if and only if

$$\begin{aligned} &D_A(f_1) \cap V(\mathfrak{a}) \neq \phi \text{ and } D_A(f_2) \cap V(\mathfrak{a}) \neq \phi \\ &\Longrightarrow D_A(f_1 \circ f_2) \cap V(\mathfrak{a}) \neq \phi \end{aligned}$$

i.e. for $\mathfrak{a} = \sqrt{\mathfrak{a}}$

$$f_1 \notin \mathfrak{a} \text{ and } f_2 \notin \mathfrak{a} \Longrightarrow f_1 \circ f_2 \notin \mathfrak{a}$$

and so $\mathfrak{a}$ is a prime. Thus under the above bijection (6.16) closed irreducible subsets of $\mathcal{spec}(A)$ correspond to the primes. Every closed irreducible subset of $\mathcal{spec}(A)$ is of the form $V(\mathfrak{p}) = \overline{\{\mathfrak{p}\}}$ for a unique prime $\mathfrak{p}$, its "generic point", and the Zariski topology is **sober**.

Proposition 6.2. *The sets $D_A(f)$, $f \in A(1)$, are compact. In particular, $\mathcal{spec}(A) = D_A(1)$ is compact.*

Proof. It is enough to show that each cover by basic open subsets has a finite subcover. We have

$$\begin{aligned} &D_A(f) \subseteq \bigcup_i D_A(g_i) \\ &\Leftrightarrow V(f) \supseteq \bigcap_i V(g_i) = V(\langle g_i, i \in I\rangle) \\ &\Leftrightarrow \sqrt{\langle f\rangle} = IV(f) \subseteq IV\{(\langle g_i\rangle)\} = \sqrt{\langle g_i\rangle} \\ &\Leftrightarrow f^m \in \langle g_i\rangle \qquad \text{for some } m \geqslant 0 \end{aligned}$$

$$\begin{aligned} &\Leftrightarrow f^m = b \circ (g_{i_k}) \circ d \quad \text{for } b \in A^-(n),\ d \in A^+(n),\ g_{i_k} \in \{g_i\}, \\ &\quad k = 1 \cdots n \\ &\vdots \\ &\Leftrightarrow D_A(f) \subseteq \bigcup_{k=1}^{n} D_A(g_{i_k}) \end{aligned}$$

□

For a homomorphism $\varphi \in \mathcal{CBio}(A, B)$, we have via pull-back

$$\mathcal{spec}(A) \longleftarrow \mathcal{spec}(B) : \varphi^*$$

It is continuous:

$$\begin{aligned} {\varphi^*}^{-1}(V_A(\mathfrak{a})) &= \{\mathfrak{q} \in \mathcal{spec}(B), \mathfrak{a} \subseteq \varphi^{-1}(\mathfrak{q})\} \\ &= \{\mathfrak{q} \in \mathcal{spec}(B), \varphi(\mathfrak{a}) \subseteq \mathfrak{q}\} = V_B(\varphi_*(\mathfrak{a})) \end{aligned} \tag{6.17}$$

Also the pull back of a **basic** open set is again a basic open set:

$$\begin{aligned}\varphi^{*^{-1}}\left(D_A(f)\right) &= \left\{\mathfrak{q} \in \mathit{spec}(B), f \notin \varphi^{-1}(\mathfrak{q})\right\} \\ &= \left\{\mathfrak{q} \in \mathit{spec}(B), \varphi(f) \notin \mathfrak{q}\right\} = D_B\left(\varphi(f)\right)\end{aligned} \tag{6.18}$$

Thus $A \rightsquigarrow \mathit{spec}(A)$ is (the object part of a contra-variant) functor from $\mathcal{CBio}$ to $\mathcal{Top}$.

Note that for $\mathfrak{b} \in \mathcal{Ideal}(B)$ we have

$$\varphi^{-1}(\sqrt{b}) = \sqrt{\varphi^{-1}(\mathfrak{b})} \tag{6.19}$$

Taking $\mathfrak{b}$ radical, $\mathfrak{b} = \sqrt{\mathfrak{b}}$, and let $\mathfrak{a} = I\left(\varphi^*\left(V(\mathfrak{b})\right)\right)$, so $V\mathfrak{a} = \overline{\varphi^*\left(V(\mathfrak{b})\right)}$, we have,

$$\begin{aligned} f \in \mathfrak{a} &\Longleftrightarrow f \in \mathfrak{p} \text{ for all } \mathfrak{p} \in \varphi^*\left(V(b)\right) \\ &\Longleftrightarrow f \in \varphi^{-1}(\mathfrak{q}) \text{ for all prime } \mathfrak{q} \supseteq \mathfrak{b}, \\ &\Longleftrightarrow \varphi(f) \in \bigcap_{\mathfrak{b} \subseteq \mathfrak{q}} q = \sqrt{\mathfrak{b}} = \mathfrak{b} \Longleftrightarrow f \in \varphi^{-1}(\mathfrak{b}) \text{ , so } \mathfrak{a} = \varphi^{-1}(\mathfrak{b}) \end{aligned}$$

Thus

$$\overline{\varphi^*\left(V(\mathfrak{b})\right)} = V\left(\varphi^{-1}\mathfrak{b}\right) \tag{6.20}$$

Chapter 7

Localization

Recall that for $A \in \mathcal{CBio}$ (respectively, $\mathcal{CProp}$), for $a \in A(1)$ (respectively, $A_{1,1}$) and $p \in A^-(n)$ or $q \in A^+(n)$ (respectively, $p \in A_{n,m}$), we write

$$\begin{aligned} a \cdot p &:= a \circ p = p \circ (a, \ldots, a) \\ a \cdot q &:= (a, \ldots, a) \circ q = q \circ a \end{aligned} \tag{7.1}$$

respectively, for prop

$$a \cdot p := \left(\bigoplus^n a\right) \circ p = p \circ \left(\bigoplus^m a\right)$$

This give an action of the monoid $A(1)$ (respectively, $A_{1,1}$) on all of A.

Given a **multiplicative subset** $S \subseteq A(1)$, (respectively, $A_{1,1}$).

$$s_1, s_2 \in S \Longrightarrow s_1 \cdot s_2 \in S; \quad 1 \in S \tag{7.2}$$

we can localize the action of S on A in the usual way. On $S \times A$ we define the equivalence relation $(s_1, p_1) \sim (s_2, p_2)$ iff

$$(s_0 \circ s_2) \cdot p_1 = (s_0 \circ s_1) \cdot p_2$$

for some $s_0 \in S$. We let $\frac{1}{s} \cdot p$ or p/s or $s^{-1} \cdot p$ denote the equivalence class of (s, p), and $S^{-1}A = (S \times A)_{/\sim}$ the collection of equivalence classes.

Using the centrality of the action we have well define operations, for $A \in \mathcal{CBio}$

$$s_0^{-1} p \circ (s_1^{-1} p_1, \ldots, s_n^{-1} p_n) := (s_0 \circ s)^{-1} \cdot p \circ ((\hat{s}_1 \cdot p_1), \ldots, \hat{s}_n \cdot p_n)$$

with

$$s = \prod_{i=1}^{n} s_i, \quad \hat{s}_i = \prod_{\substack{j=1 \\ j \neq i}}^{n} s_j$$

and similarly for left or right actions.

For $A \in \mathcal{CProp}$, we have well defined operations

$$\begin{aligned}(s_0^{-1}p_0) \circ (s_1^{-1}p_1) &:= (s_0 \cdot s_1)^{-1}(p_0 \circ p_1)\\ (s_0^{-1}p_0) \oplus (s_1^{-1}p_1) &:= (s_0 \cdot s_1)^{-1}(s_1 \cdot p_0 \oplus s_0 \cdot p_1)\end{aligned}$$

These operations makes the collection $S^{-1}A$ of equivalence classes into a commutative bio (respectively, prop), and the **canonical homomorphism**

$$\rho_S : A \longrightarrow S^{-1}A, \quad \rho_S(a) = a/1 \tag{7.3}$$

is universal with respect to inverting S

$$\mathcal{CBio}(S^{-1}A, B) \equiv \{\varphi \in \mathcal{CBio}(A, B), \varphi_1(S) \subseteq \mathrm{GL}_1(B)\} \tag{7.4}$$

and similarly for props. The functor

$$\begin{aligned}&\mathcal{CBio}_{A\backslash} \longrightarrow \mathcal{CBio}_{S^{-1}A\backslash}\\ &(A \xrightarrow{\varphi} B) \rightsquigarrow \left(S^{-1}A \xrightarrow{S^{-1}\varphi} \varphi(S)^{-1}B\right)\end{aligned} \tag{7.5}$$

commutes with arbitrary co-limits, and (since we can always take a "common-denominator") also commute with finite limits.

We say $\mathfrak{a} \in \mathcal{Ideal}(A)$ is ***S*-saturated** if

$$s \in S\ ,\ s \circ a \in \mathfrak{a} \Longrightarrow a \in \mathfrak{a}$$

Proposition 7.1. *We have the bijection*

$$\{\mathfrak{a} \in \mathcal{Ideal}(A),\ \mathfrak{a}\ S\text{-saturated}\} \xleftrightarrow{\sim} \mathcal{Ideal}\,(S^{-1}A)$$

Proof. We have $(\rho_S)_* \circ \rho_S^* = \mathrm{id}$. Indeed, for $\mathfrak{b} \in \mathcal{Ideal}(S^{-1}A)$, if $s^{-1} \cdot a \in \mathfrak{b}$, then $a/1 \in \mathfrak{b}$, $a \in \rho_S^*(\mathfrak{b})$, $s^{-1} \cdot a \in (\rho_S)_*\rho_S^*(\mathfrak{b})$. Thus

$\mathfrak{b} \subseteq (\rho_S)_* \rho_S^*(\mathfrak{b})$, and since the opposite inclusion always holds we get equality:

$$\mathfrak{b} \equiv (\rho_S)_* \, \rho_S^*(\mathfrak{b}) \equiv S^{-1} \rho_S^*(\mathfrak{b}) := \{s^{-1} \cdot a, s \in S, a \in \rho_S^*(\mathfrak{b})\}$$

For $\mathfrak{b} \in \mathcal{I}deal(S^{-1}A)$, clearly $\rho_S^*(\mathfrak{b})$ is S-saturated.

Let $\mathfrak{a} \in \mathcal{I}deal(A)$ be S-saturated, and let $c \in \rho_S^*(\rho_S)_*(\mathfrak{a})$, so

$$c/1 = s_1^{-1} b \circ (a_i) \circ d/s_2, \quad s_i \in S, \quad b \in A^-(n), \quad d \in A^+(n), \quad a_i \in \mathfrak{a}$$

so for some $s_0 \in S$.

$$s_0 \circ s_1 \circ s_2 \circ c = s_0 \circ b \circ (a_i) \circ d \in \mathfrak{a} \xRightarrow{S-\text{saturated}} c \in \mathfrak{a}$$

Thus $\rho_S^*(\rho_S)_*(\mathfrak{a}) \subseteq \mathfrak{a}$, and since the opposite inclusion always hold, we have an equality.

Thus the image of ρ_S^* is exactly the set of S-saturated ideals of A. □

For any ideal $\mathfrak{a} \in \mathcal{I}deal(\mathrm{A})$, the ideal $\rho_S^*(\rho_S)_*(\mathfrak{a})$ is the S-saturation of $\mathfrak{a}$, and it can be written as the directed union of ideals

$$\begin{aligned} &\rho_S^*(\rho_S)_*(\mathfrak{a}) = \bigcup_{s \in S} (\mathfrak{a} : s) \\ &(\mathfrak{a} : s) := \{x \in A(1), x \circ s \in \mathfrak{a}\} \\ &(\mathfrak{a} : s_1) \cup (\mathfrak{a} : s_2) \subseteq (\mathfrak{a} : s_1 \circ s_2) \end{aligned} \tag{7.6}$$

Corollary 7.2. *We get the induced bijection*

$$\{\mathfrak{p} \in spec(A), \mathfrak{p} \cap S = \varnothing\} \xleftrightarrow{\sim} spec(S^{-1}A)$$

$$\{a \in A, a/1 \in \mathfrak{q}\} \equiv \rho_S^*(\mathfrak{q}) \longleftarrow\!\shortmid \mathfrak{q}$$

$$\mathfrak{p} \longmapsto (\rho_S)_*(\mathfrak{p}) \equiv S^{-1}\mathfrak{p}$$

We will be interested in the following two cases.

First for $f \in A(1)$ we have: $S_f = f^{\mathbb{N}} = \{1, f, f^2, \ldots, f^n, \ldots\}$.

We write $A\,[1/f] := S_f^{-1}A$.

$$spec(A) \supseteq D_A(f) \xleftrightarrow{\sim} spec\,(A\,[1/f]) \tag{7.7}$$

Secondly, for a prime $\mathfrak{p} \in spec(A)$, $S_\mathfrak{p} = A(1) \backslash \mathfrak{p}$.

We write $A_\mathfrak{p} := S_\mathfrak{p}^{-1}A$.

$$\{\mathfrak{q} \in spec(A), \mathfrak{q} \subseteq \mathfrak{p}\} \xleftrightarrow{\sim} spec(A_\mathfrak{p}) \tag{7.8}$$

In particular, $A_{\mathfrak{p}}$ is a **local** bio (respectively, prop) in that it has a **unique maximal ideal**

$$\{\mathfrak{m}_{\mathfrak{p}} \equiv S_{\mathfrak{p}}^{-1}(\mathfrak{p})\} \equiv max(A_{\mathfrak{p}}) \tag{7.9}$$

Chapter 8

Sheaves

Fix $P \in \mathcal{CBio}$ (respectively, $\mathcal{CProp}$). For an open set $\mathcal{U} \subseteq \mathit{spec}(P)$ define

$$\mathcal{O}_P^{\pm}(\mathcal{U})(n) := \left\{ f : \mathcal{U} \to \coprod_{\mathfrak{p} \in \mathcal{U}} P_{\mathfrak{p}}^{\pm}(n), f(\mathfrak{p}) \in P_{\mathfrak{p}}^{\pm}(n), \right.$$

$$\left. \text{and } f \text{ is a } \textbf{local fraction} \right\} \tag{8.1}$$

f **locally a fraction**: for all $\mathfrak{p} \in \mathcal{U}$, there exists open set $\mathcal{U}_{\mathfrak{p}}$, $\mathfrak{p} \in \mathcal{U}_{\mathfrak{p}} \subseteq \mathcal{U}$, and $p \in P^{\pm}(n)$, $s \in P(1) \setminus \bigcup_{\mathfrak{q} \in \mathcal{U}_{\mathfrak{p}}} \mathfrak{q}$, such that for all $\mathfrak{q} \in \mathcal{U}_{\mathfrak{p}}$ we have $f(\mathfrak{q}) \equiv p/s$ in $P_{\mathfrak{q}}$.

Respectively, for a prop $P \in \mathcal{CProp}$, we define $\mathcal{O}_P(\mathcal{U})_{n,m}$ as the local fractions taking values in $(P_{\mathfrak{p}})_{n,m}$.

Note that $\mathcal{O}_P(\mathcal{U}) = \left(\mathcal{O}_P^-(\mathcal{U}), \mathcal{O}_P^+(\mathcal{U})\right)$ (respectively, $\{\mathcal{O}_P(\mathcal{U})_{n,m}\}$) is a commutative bio (respectively, prop) the operations of compositions and actions are defined pointwise in each $P_{\mathfrak{p}}$, and the "local fraction condition" is preserved.

For open sets $\mathcal{V} \subseteq \mathcal{U} \subseteq \mathit{spec}(P)$ we have the restriction maps

$$\rho_{\mathcal{V}}^{\mathcal{U}} : \mathcal{O}_P(\mathcal{U}) \to \mathcal{O}_P(\mathcal{V}) \tag{8.2}$$

making $\mathcal{U} \mapsto \mathcal{O}_P(\mathcal{U})$ a pre-sheaf of bios (respectively, props) over $\mathit{spec}(P)$.

By the local nature of the "locally-fraction-condition" it is clearly a sheaf, and each $\mathcal{O}_P^{\pm}(n)$ (respectively $(\mathcal{O}_P)_{n,m}$) is a sheaf of sets.

For $\mathfrak{p} \in \mathit{spec}(P)$, the **stalk** of $\mathcal{O}_P$ at $\mathfrak{p}$ is given by

$$\mathcal{O}_{P,\mathfrak{p}} := \varinjlim_{\mathcal{U} \ni \mathfrak{p}} \mathcal{O}_P(\mathcal{U}) \xrightarrow{\sim} P_{\mathfrak{p}} \tag{8.3}$$

via taking the value of a section at the point $\mathfrak{p}$

$$(\mathcal{U}, f)_{/\approx} \longmapsto f(\mathfrak{p})$$

It is well defined, surjective, and injective.

The global sections of $\mathcal{O}_P$ are given by

Theorem 8.1. *For $P \in \mathcal{CB}io$ or $P \in \mathcal{CP}rop$. For a basic open set $D(s) \subseteq \mathit{spec}(P)$, $s \in P(1) \equiv P_{1,1}$, we have bijection*

$$\Psi : P_s = P\left[\tfrac{1}{s}\right] \xrightarrow{\quad\sim\quad} \mathcal{O}_P(D(s))$$
$$\Psi\left(p/s^n\right) := \{f(\mathfrak{p}) \equiv p/s^n \text{ in } P_{\mathfrak{p}} \text{ for all } \mathfrak{p} \in D(s)\}$$

For $s = 1$ we obtain the global sections

$$P \xrightarrow{\quad\sim\quad} \mathcal{O}_P(\mathit{spec}(P))$$

Proof. The map Ψ which takes $p/s^n \in P_s$ to the constant section f, $f(\mathfrak{p}) \equiv p/s^n$, is a well defined map of bios (respectively, props).

Take $p \in P^-(k)$, the case $p \in P^+(k)$ is similar (respectively, take $p \in P_{n,m}$).

Ψ is injective: Assume $\Psi\left(p_1/s^{n_1}\right) = \Psi(p_2/s^{n_2})$, and define

$$\mathfrak{a} := \mathrm{Ann}\left(s^{n_2} \circ p_1; s^{n_1} \circ p_2\right) = \{a \in P(1), a \circ s^{n_2} \circ p_1 = a \circ s^{n_1} \circ p_2\}$$

by commutativity it follows $\mathfrak{a}$ is an ideal. We have

$$p_1/s^{n_1} = p_2/s^{n_2} \text{ in } P_{\mathfrak{p}} \text{ for all } \mathfrak{p} \in D(s)$$

$\Rightarrow\ s_{\mathfrak{p}} \circ s^{n_2} \circ p_1 = s_{\mathfrak{p}} \circ s^{n_1} \circ p_2$ for some $s_{\mathfrak{p}} \in P(1)\backslash\mathfrak{p}$, all $\mathfrak{p} \in D(s)$
$\Rightarrow\ \mathfrak{a} \nsubseteq \mathfrak{p}$, all $\mathfrak{p} \in D(s)$
$\Rightarrow\ V(\mathfrak{a}) \cap D(s) = \varnothing$
$\Rightarrow\ V(\mathfrak{a}) \subseteq V(s)$
$\Rightarrow\ s \in IV(\mathfrak{a}) = \sqrt{\mathfrak{a}}$
$\Rightarrow\ s^n \in \mathfrak{a}$ for $n \gg 0$
$\Rightarrow\ s^{n+n_2} \circ p_1 = s^{n+n_1} \circ p_2$ for $n \gg 0$
$\Rightarrow\ p_1/s^{n_1} = p_2/s^{n_2}$ in P_s

Ψ is surjective: Fix $f \in \mathcal{O}_P(D(s))^-(k)$, respectively, $\mathcal{O}_P(D(s))_{k_1,k_2}$.

Since $D(s)$ is compact we can cover it by a finite collection of basic open sets

$$D(s) = D(s_1) \cup \cdots \cup D(s_N)$$

with

$$f(\mathfrak{p}) \equiv p_i/t_i \text{ for } \mathfrak{p} \in D(s_i); \quad t_i \in P(1) \setminus \bigcup_{\mathfrak{p} \in D(s_i)} \mathfrak{p}$$

We have $V(t_i) \subseteq V(s_i)$, so $s_i^{n_i} = c_i \circ t_i$ for some $c_i \in P(1)$, and for $\mathfrak{p} \in D(s_i)$, $f(\mathfrak{p}) = p_i/t_i = c_i \cdot p_i/s_i^{n_i}$.

We can replace s_i by $s_i^{n_i}$, $(D(s_i) = D(s_i^{n_i}))$, and p_i by $c_i \cdot p_i$, so

$$f(\mathfrak{p}) \equiv p_i/s_i \quad \text{for } \mathfrak{p} \in D(s_i)$$

On the set $D(s_i) \cap D(s_j) = D(s_i \circ s_j)$, the section f is given by both p_i/s_i and p_j/s_j, and by the injectivity of Ψ

$$(s_i \circ s_j)^n \circ s_j \cdot p_i = (s_i \circ s_j)^n \circ s_i \cdot p_j$$

By finiteness we may assume one n works for all i, j.

Replacing s_i by s_i^{n+1}, and replacing $s_i^n \cdot p_i$ by p_i, we may assume

$$\begin{aligned} &f(\mathfrak{p}) \equiv p_i/s_i \quad \text{for all } \mathfrak{p} \in D(s_i) \\ &s_j \cdot p_i = s_i \cdot p_j \quad \text{all } i, j \end{aligned}$$

Since $D(s) \subseteq \bigcup_i D(s_i)$ we have that some power s^M is a linear-combination of the s_i

$$\begin{aligned} &s^M = b \circ (c_j) \circ d, \quad (\text{respectively, } b \circ (\oplus c_j) \circ d), \\ &c_j = s_{i(j)}, \quad b, d \in P^{\pm}(\ell), \\ &j \in \{1, \ldots, \ell\} \xrightarrow{i} \{1, \ldots, N\}. \end{aligned}$$

Define $p = b \circ (p_{i(j)}) \circ \sigma_{\ell,k} \overleftarrow{\circ} (\underbrace{d, \ldots, d}_{k})$

$$\text{resp., } p = \left(\bigoplus^{k_1} b \right) \circ \sigma_{k_1,\ell} \circ \left(\bigoplus_{j=1}^{\ell} p_{i(j)} \right) \circ \sigma_{\ell,k_2} \circ \left(\bigoplus^{k_2} d \right)$$

We have, for any $j_0 \in \{1, \ldots, N\}$,

$$\begin{aligned} s_{j_0} \circ p &= b \circ (s_{j_0} \circ p_{i(j)}) \circ \sigma_{\ell,k} \overleftarrow{\circ} (d, \ldots, d) \\ &\equiv b \circ (s_{i(j)} \circ p_{j_0}) \circ \sigma_{\ell,k} \overleftarrow{\circ} (d, \ldots, d) \end{aligned}$$

$$= (b \circ (c_j \circ p_{j_0})) \circ \sigma_{\ell,k} \overleftarrow{\circ} (d, \dots, d)$$

$$= ((b \circ (c_j)) \overleftarrow{\circ} d) \circ p_{j_0} \quad \text{by commutativity}$$

$$= s^M \circ p_{j_0}$$

Respectively for a commutative prop,

$$s_{j_0} \cdot p = \left(\bigoplus^{k_1} b \right) \circ \sigma_{k_1,\ell} \circ \left(\bigoplus_{j=1}^{\ell} s_{j_0} \cdot p_{i(j)} \right) \circ \sigma_{\ell,k_2} \circ \left(\bigoplus^{k_2} d \right)$$

$$= \left(\bigoplus^{k_1} b \right) \circ \sigma_{k_1,\ell} \circ \left(\bigoplus_{j=1}^{\ell} s_{i(j)} \cdot p_{j_0} \right) \circ \sigma_{\ell,k_2} \circ \left(\bigoplus^{k_2} d \right)$$

$$= \left(b \circ \left(\bigoplus_{j=1}^{\ell} s_{i(j)} \right) \circ d \right) \cdot p_{j_0} \quad \text{by commutativity}$$

$$= s^M \cdot p_{j_0}$$

Thus $f(\mathfrak{p}) \equiv p_{j_0}/s_{j_0} \equiv p/s^M$ is constant. □

Chapter 9

Generalized Schemes

For a topological space $\mathcal{X}$, we have the category $\mathcal{CBio}/\mathcal{X}$ of sheaves of $\mathcal{CBio}$ over $\mathcal{X}$, with maps the natural transformations $\varphi = \{\varphi_{\mathcal{U}}\}$, $\varphi_{\mathcal{U}} \in \mathcal{CBio}(P(\mathcal{U}), P'(\mathcal{U}))$. Putting all these categories together we have the category $\mathcal{CBio}/\mathcal{Top}$: its object are pairs $(\mathcal{X}, P)$, $\mathcal{X} \in \mathcal{Top}$, $P \in \mathcal{CBio}/\mathcal{X}$, and its maps $f : (\mathcal{X}, P) \to (\mathcal{X}', P')$ are pairs $f \in \mathcal{Top}(\mathcal{X}, \mathcal{X}')$ and $f^\natural \in \mathcal{CBio}/\mathcal{X}'(P', f_*P)$; explicitly, f is a continuous function, and for $\mathcal{U} \subseteq \mathcal{X}'$ open, we have the map of bios $f^\natural_{\mathcal{U}} \in \mathcal{CBio}\left(P'(\mathcal{U}), P\left(f^{-1}(\mathcal{U})\right)\right)$, these maps being compatible with restrictions. Similarly, we have the category $\mathcal{CProp}/\mathcal{Top}$ with objects the pairs $(\mathcal{X}, P)$, $\mathcal{X}$ a topological space and P a sheaf of commutative props over $\mathcal{X}$.

Remark 9.1. For $f \in \mathcal{Top}(\mathcal{X}, \mathcal{X}')$ we have adjunctions

$$\begin{array}{ccc} \mathcal{CBio}/\mathcal{X} & & \mathcal{CProp}/\mathcal{X} \\ f^* \uparrow \;\; \downarrow f_* & \text{and} & f^* \uparrow \;\; \downarrow f_* \\ \mathcal{CBio}/\mathcal{X}' & & \mathcal{CProp}/\mathcal{X}' \end{array}$$

$$f^*P' = \left\{ \begin{array}{c} \text{sheaf associated to the pre-sheaf} \\ \mathcal{X} \supseteq \mathcal{V} \longmapsto \varinjlim P'(\mathcal{U}) \\ \mathcal{U} \subseteq \mathcal{X}' \text{ open} \\ f(\mathcal{V}) \subseteq \mathcal{U} \end{array} \right\} \qquad f_*P(\mathcal{U}) = P(f^{-1}\mathcal{U})$$

For a map $f \in \mathcal{CB}io/\mathcal{T}op\left((\mathcal{X},P),(\mathcal{X}',P')\right)$, and for a point $x \in \mathcal{X}$, we get an induced map of stalks

$$f_x^\natural : P'_{f(x)} \equiv \varinjlim_{f(x)\in\mathcal{V}\subseteq\mathcal{X}'} P'(\mathcal{V}) \longrightarrow \varinjlim_{f(x)\in\mathcal{V}\subseteq\mathcal{X}'} P(f^{-1}\mathcal{V}) \longrightarrow \varinjlim_{x\in\mathcal{U}\subseteq\mathcal{X}} P(\mathcal{U}) \equiv P_x \tag{9.1}$$

Definition 9.1. The category of **locally-bio-spaces** $\mathcal{CB}io_{\text{loc}}/\mathcal{T}op$ (respectively, **locally-prop-space** $\mathcal{CP}rop_{\text{loc}}/\mathcal{T}op$), is the category with object $(\mathcal{X},P) \in \mathcal{CB}io/\mathcal{T}op$, respectively, $\mathcal{CP}rop/\mathcal{T}op$, such that for all $x \in \mathcal{X}$ the stalk P_x is a **local** bio (respectively, prop) with the unique maximal ideal $\mathfrak{m}_x \subseteq P_x(1)$; the maps $f \in \mathcal{CB}io_{\text{loc}}/\mathcal{T}op\left((\mathcal{X},P),(\mathcal{X}',P')\right)$ are maps $f \in \mathcal{CB}io/\mathcal{T}op\left((\mathcal{X},P),(\mathcal{X}',P')\right)$ such that for all $x \in X$,

$$f_x^\natural \in \mathcal{CB}io_{\text{loc}}\left(P'_{f(x)}, P_x\right) \quad \text{respectively,} \quad \mathcal{CP}rop_{\text{loc}}(P'_{f(x)}, P_x)$$

is a **local** map: $f_x^\natural(\mathfrak{m}_{f(x)}) \subseteq \mathfrak{m}_x$.

Theorem 9.1. *We have the adjunction*

$$\begin{array}{c} \left(\mathcal{CB}io_{\text{loc}}/\mathcal{T}op\right)^{\text{op}} \\ \mathit{spec} \Big\uparrow \quad \Big\downarrow \Gamma \\ \mathcal{CB}io \end{array}$$

$$\mathit{spec}(P) := \left(\mathit{spec}(P), \mathcal{O}_P\right) \qquad \Gamma(\mathcal{X},P) := P(\mathcal{X}) \ \text{the global sections}$$

$$\mathcal{CB}io_{loc}/\mathcal{T}op\left((\mathcal{X},P),\mathit{spec}(A)\right) \equiv \mathcal{CB}io\left(A, P(\mathcal{X})\right)$$

respectively,

$$\mathcal{CP}rop_{loc}/\mathcal{T}op\left((\mathcal{X},P),\mathit{spec}(A)\right) \equiv \mathcal{CP}rop\left(A, P(\mathcal{X})\right)$$

Proof. We work with bios, the case of props is similar. For a point $x \in \mathcal{X}$, we have canonical map $\phi_x \in \mathcal{CB}io\left(P(\mathcal{X}), P_x\right)$, $\phi_x a = a|_x$ the stalk of the global section a at the point x, and we get a prime $\mathfrak{p}_x := \phi_x^{-1}(\mathfrak{m}_x) \in \mathit{spec}\left(P(\mathcal{X})\right)$, $\mathfrak{m}_x \subseteq P_x(1)$ the maximal ideal.

For a basic open set $D(s) \subseteq spec\,(P(\mathcal{X}))$, $s \in P(\mathcal{X})(1)$, we have

$$\{x \in \mathcal{X}, \mathfrak{p}_x \in D(s)\} = \{x \in \mathcal{X}, \phi_x(s) \notin \mathfrak{m}_x\} \text{ is } \textbf{open} \text{ in } \mathcal{X}:$$

if $\phi_x(s) \notin \mathfrak{m}_x$ it is invertible in P_x, $a_x \circ \phi_x(s) = 1$, $a_x \in P_x$ and there is an open $\mathcal{U}_x \ni x$, $a \in P(\mathcal{U}_x)$ with $a_x = a|_x$; taking $\mathcal{U}_x$ smaller we have $a \circ s|_{\mathcal{U}_x} = 1$ already in $P(\mathcal{U}_x)$, and so $\phi_{x'}(s) \notin m_{x'}$ for all $x' \in \mathcal{U}_x$.

Thus the map $x \mapsto \mathfrak{p}_x$ is a continuous map $\mathfrak{p} : \mathcal{X} \longmapsto spec\,(P(\mathcal{X}))$.

The uniqueness of the inverse $(s|_{u_x})^{-1}$, shows these local inverses glue to a global inverse $s^{-1} \in P\left(\mathfrak{p}^{-1}(D(s))\right)$ and we get the map of $\mathcal{CB}io$

$$\mathfrak{p}^{\natural}_{D(s)} : P(\mathcal{X})_s \equiv \left\{s^{\mathbb{N}}\right\}^{-1} P(\mathcal{X}) \longrightarrow P\left(\mathfrak{p}^{-1}(D(s))\right) \tag{9.2}$$

These maps are compatible on intersections,

$$D(s_1) \cap D(s_2) = D(s_1 \circ s_2),$$

so the sheaf property give $\mathfrak{p}^{\natural}_{\mathcal{U}} \in \mathcal{CB}io(\mathcal{O}_{spec\,P(\mathcal{X})}(\mathcal{U}), P(\mathfrak{p}^{-1}\mathcal{U}))$ for any open $\mathcal{U} \subseteq spec\,(P(\mathcal{X}))$, compatible with restrictions, so

$$\mathfrak{p} = (\mathfrak{p}, \mathfrak{p}^{\natural}) \in \mathcal{CB}io/\mathcal{T}op\left((\mathcal{X}, P), spec\,(P(\mathcal{X}))\right) \tag{9.3}$$

For $x \in \mathcal{X}$, we get

$$\mathfrak{p}^{\natural}_x = \varinjlim_{\phi_x(s) \notin \mathfrak{m}_x} \mathfrak{p}^{\natural}_{D(s)} \in \mathcal{CB}io_{\text{loc}}\,(P(\mathcal{X})_{\mathfrak{p}_x}, P_x) \tag{9.4}$$

is **local**, so $\mathfrak{p} \in \mathcal{CB}io_{\text{loc}}/\mathcal{T}op\left((\mathcal{X}, P), spec\,(P(\mathcal{X}))\right)$; it is the co-unit of adjunction.

Given $\varphi \in \mathcal{CB}io(A, P(\mathcal{X}))$ we get

$$spec(\varphi) \circ \mathfrak{p} \in \mathcal{CB}io_{\text{loc}}/\mathcal{T}op((\mathcal{X}, P), spec(A)).$$

Given $f \in \mathcal{CB}io_{\text{loc}}/\mathcal{T}op\left((\mathcal{X}, P), spec(A)\right)$ we get the global sections

$$\Gamma(f) = f^{\natural}_{spec(A)} \in \mathcal{CB}io(A, P(\mathcal{X})). \tag{9.5}$$

One checks these are inverse bijections using the locality of f. $\square$

Following the footsteps of Grothendieck we can define the categories of generalized schemes, as the locally-bio or prop-spaces that are "locally-affine".

Definition 9.2. Bio Schemes $\mathcal{BSch} \subseteq \mathcal{CBio}_{\text{loc}}/\mathcal{Top}$ is the full subcategory of $\mathcal{CBio}_{\text{loc}}/\mathcal{Top}$ consisting of the object $(\mathcal{X}, \mathcal{O}_{\mathcal{X}})$ which are locally affine: we have some open cover $\mathcal{X} = \bigcup \mathcal{U}_i$, and

$$(\mathcal{U}_i, \mathcal{O}_{\mathcal{X}}|_{\mathcal{U}_i}) \cong \mathit{spec}\,(\mathcal{O}_{\mathcal{X}}(\mathcal{U}_i))$$

Similarly, we have the full subcategory $\mathcal{PSch} \subseteq \mathcal{CProp}_{\text{loc}}/\mathcal{Top}$ of **Prop Schemes** with objects the locally affine prop-spaces.

Insisting that all our props (respectively, bios) are totally-commutative (respectively, fully-commutative) we get the full subcategories of generalized schemes

$$P_T\mathcal{Sch} \subseteq P_f\mathcal{Sch} \subseteq \mathcal{PSch}$$

respectively,

$$B_T\mathcal{Sch} \subseteq B_f\mathcal{Sch} \subseteq \mathcal{BSch}$$

Proposition 9.2. *An open subset of a generalized scheme is again a generalized scheme.*

Proof. This follows since we have the basis $D_A(f)$ for $\mathit{spec}A$, and $D_A(f) \equiv \mathit{spec}A[1/f]$ is affine. □

Proposition 9.3. *Schemes can be glued along open subsets and consistent glueing data.*

Thus if we have (generalized) schemes $X_i, i \in I$, and open subsets $\mathcal{U}_{i_j} \subseteq X_i$, and isomorphisms of schemes $\phi_{j_i} : \mathcal{U}_{i_j} \xrightarrow{\sim} \mathcal{U}_{j_i}$, satisfying

$$\begin{gathered} \mathcal{U}_{i_i} = X_i, \quad \phi_{i_i} = \text{id}_{X_i}\ , \quad \text{all } i \in I \\ \phi_{k_i} = \phi_{k_j} \circ \phi_{j_i} \text{ on } \mathcal{U}_{i_j} \cap \mathcal{U}_{i_k} \text{ all } i, j, k \in I \end{gathered} \tag{9.6}$$

there exists a unique (generalized) scheme X, and open cover $X = \bigcup_{i \in I} \mathcal{U}_i$, and isomorphisms of schemes $\psi_i : X_i \xrightarrow{\sim} \mathcal{U}_i$ for all $i \in I$, such that for $i, j \in I$

$$\begin{array}{ccc} & \mathcal{U}_i \cap \mathcal{U}_j & \\ {\scriptstyle \psi_i}\nearrow\wr & & \wr\nwarrow{\scriptstyle \psi_j} \\ X_i \supseteq \mathcal{U}_{i_j} & \underset{\sim}{\overset{\phi_{i_j}}{\longleftarrow}} & \mathcal{U}_{j_i} \subseteq X_j \end{array} \qquad \text{is commutative} \tag{9.7}$$

Since ordinary commutative rings A give totally-commutative bios $\mathcal{C}_T\mathcal{B}io$ (respectively, $\mathcal{C}_T\mathcal{P}rop$) and since all our definitions reduce to their classical analogues for $A \in \mathcal{CR}ing$, ordinary schemes embeds fully faithfully in generalized schemes, and we have the full and faithful embeddings over $\mathcal{T}op$:

$$\begin{array}{ccccccc} & & P_T\mathcal{S}ch & \hookrightarrow & P_f\mathcal{S}ch & \hookrightarrow & \mathcal{PS}ch \\ & \nearrow & \downarrow\uparrow & & \downarrow\uparrow & & \downarrow\uparrow \\ \mathcal{S}ch & & & & & & \\ & \searrow & & & & & \\ & & B_T\mathcal{S}ch & \hookrightarrow & B_f\mathcal{S}ch & \hookrightarrow & \mathcal{BS}ch \end{array} \tag{9.8}$$

Theorem 9.4. *The categories $\mathcal{PS}ch$ and $\mathcal{BS}ch$, and their full subcategories $P_T\mathcal{S}ch \subseteq P_f\mathcal{S}ch$ and $B_T\mathcal{S}ch \subseteq B_f\mathcal{S}ch$, have fiber products, for $f_i \in \mathcal{BS}ch(X_i, Y)$ or $\mathcal{PS}ch(X_i, Y)$:*

$$\begin{array}{ccccc} & & X_0 \prod_Y X_1 & & \\ & \swarrow & & \searrow & \\ X_0 & & & & X_1 \\ & {\scriptstyle f_0}\searrow & & \swarrow{\scriptstyle f_1} & \\ & & Y & & \end{array}$$

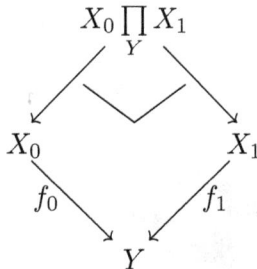

Proof. Exactly as for ordinary schemes. Write $Y = \bigcup_i \mathcal{spec}(B_i)$, $X_\varepsilon = \bigcup_{i,j} \mathcal{spec}(A^\varepsilon_{j,i})$, $\varepsilon = 0, 1$ with $f_\varepsilon(\mathcal{spec}(A^\varepsilon_{j,i})) \subseteq \mathcal{spec}\ (B_i)$, and glue:

$$X_0 \prod_Y X_1 \equiv \coprod_{i,j_0,j_1} \mathcal{spec}\left(A^0_{j_0,i} \underset{B_i}{\boxtimes} A^1_{j_1,i}\right) \tag{9.9}$$

Note that one uses the push-out $(A^0_{j_0,i} \coprod_{B_i} A^1_{j_1,i})^C$ in $\mathcal{CBio}$ or $\mathcal{CProp}$ for $\mathcal{BSch}$ or $\mathcal{PSch}$, and its totally-commutative (respectively, fully-commutative) quotient for the appropriate full subcategory, i.e. $A^0_{j_0,i} \boxtimes_{B_i}^T A^1_{j_1,i}$ (respectively, $A^0_{j,i} \boxtimes_{B_i}^f A^1_{j_1,i}$) in the totally (respectively, fully)-commutative case. □

Chapter 10

Pro-Schemes

We have the full-embeddings

$$\mathcal{BAff} \equiv (\mathcal{CBio})^{\mathrm{op}} \subseteq \mathcal{BSch} \subseteq \mathcal{CBio}_{\mathrm{loc}}/\mathcal{Top} \tag{10.1}$$

respectively,

$$\mathcal{PAff} \equiv (\mathcal{CProp})^{\mathrm{op}} \subseteq \mathcal{PSch} \subseteq \mathcal{CProp}_{\mathrm{loc}}/\mathcal{Top}$$

The categories $\mathcal{CBio}_{\mathrm{loc}}/\mathcal{Top}$ and $\mathcal{CProp}_{\mathrm{loc}}/\mathcal{Top}$ are "complete":

Given a functor $I \ni i \rightsquigarrow \mathcal{X}_i$ from a partially ordered set I, we have the topological space $\mathcal{X} = \lim\limits_{\overleftarrow{I}} \mathcal{X}_i$, its points $x = (x_i)$ are coherent sequences of points $x_i \in \mathcal{X}_i$ and we give it the inverse limits topology: the smallest topology such that all the projections $\pi_i : \mathcal{X} \to \mathcal{X}_i$ are continuous. We have the structure sheaf $\mathcal{O}_{\mathcal{X}} = \lim\limits_{\overrightarrow{I}} \pi_i^* \mathcal{O}_{\mathcal{X}_i}$ over it, and it has local stalks, indeed:

$$\mathcal{O}_{\mathcal{X}}\Big|_{(x_i)} \equiv \lim_{\substack{\longrightarrow \\ i \in I}} \mathcal{O}_{\mathcal{X}_i, x_i}$$

the co-limit of local homomorphisms is local again.

The categories $\mathcal{BAff}$ (respectively, $\mathcal{PAff}$) are also complete, as $\mathcal{CBio}$ (respectively, $\mathcal{CProp}$) is co-complete and

$$\lim_{\overleftarrow{I}} \mathit{spec}(A_i) \equiv \mathit{spec}(\lim_{\overrightarrow{I}} A_i) \tag{10.2}$$

But the categories in the middle $\mathcal{BSch}$ or $\mathcal{PSch}$, the full sub-categories of $\mathcal{CBio}_{\mathrm{loc}}/\mathcal{Top}$ or $\mathcal{CProp}_{\mathrm{loc}}/\mathcal{Top}$ with objects the ones locally isomorphic to elements of $\mathcal{Aff}$, are not complete: Given functor $I \ni i \rightsquigarrow \mathcal{X}_i$

into (generalized) scheme and given a point

$$x = (x_i) \in \lim_{\overleftarrow{I}} \mathcal{X}_i \tag{10.3}$$

while for each $i \in I$ we have an affine neighborhood $x_i \in \mathit{spec}A_i \subseteq \mathcal{X}$ there need not be an affine neighborhood of x in $\lim_{\overleftarrow{I}} \mathcal{X}_i$ (such is the case of the "real primes" in Example 10.1). We have exactly a similar discussion for the categories $\mathcal{B}_T\mathcal{S}ch \subseteq \mathcal{B}_f\mathcal{S}ch$ (respectively, $\mathcal{P}_T\mathcal{S}ch \subseteq \mathcal{P}_f\mathcal{S}ch$): they are the full-subcategories of the complete categories

$$\mathcal{C}_T\mathcal{B}io_{\text{loc}}/\mathcal{T}op \subseteq \mathcal{C}_f\mathcal{B}io_{\text{loc}}/\mathcal{T}op$$

(respectively, $\mathcal{C}_T\mathcal{P}rop_{\text{loc}}/\mathcal{T}op \subseteq \mathcal{C}_f\mathcal{P}rop_{\text{loc}}/\mathcal{T}op$) which are locally affine, and these categories of generalized schemes are not complete.

So we pass to the pro-categories. It has objects $(J, \mathcal{X}_j)$ where J is a partially ordered set which we always take to be

$$\text{directed:} \quad \forall j_1, j_2 \in J, \quad \exists j \in J, \quad j \geqslant j_1 \quad \text{and} \quad j \geqslant j_2 \tag{10.4}$$

$$\text{co-finite:} \quad \forall j \in J, \quad \#\{i \in J, i \leqslant j\} < \infty \tag{10.5}$$

and

$$I \ni j \rightsquigarrow \mathcal{X}_j \text{ is a functor :} \quad j_1 \geqslant j_0 \implies \mathcal{X}_{j_1} \longrightarrow \mathcal{X}_{j_0}$$

We shall use $\mathcal{G}$ for $\mathcal{GS}ch$ "generalized" schemes, for either $\mathcal{G} = \mathcal{B}$, $\mathcal{BS}ch = \mathcal{B}io$-schemes, or $\mathcal{G} = \mathcal{B}_T$, or $\mathcal{G} = \mathcal{B}_f$, or $\mathcal{G} = \mathcal{P}$, $\mathcal{PS}ch = \mathcal{P}rop$ schemes or $\mathcal{G} = \mathcal{P}_T$ or $\mathcal{G} = \mathcal{P}_f$. The set of maps in pro-$\mathcal{GS}ch$ are given by

$$\mathit{pro}\text{-}\mathcal{GS}ch((J, X_j), (I, Y_i)) = \lim_{\overleftarrow{I}} \lim_{\overrightarrow{J}} \mathcal{GS}ch(X_j, Y_i) \tag{10.6}$$

We can describe this set as the collection of $\varphi = \{\varphi_{i_j}\}$,

$$\varphi_{i_j} \in \mathcal{GS}ch(X_j, Y_i) \quad \text{for all } i \in I, \text{ and for } j \geqslant \sigma(i) \text{ large}$$

satisfying for $i' \leqslant i$ and $\sigma(i) \leqslant j \leqslant j'$

$$\varphi_{i_j} \circ \pi^X_{j_{j'}} = \varphi_{i_{j'}}$$

$$\pi^Y_{i'_i} \circ \varphi_{i_j} = \varphi_{i'_j}$$

(where $\pi^X_{j_{j'}} \in \mathcal{GSch}(X_{j'}, X_j)$, $\pi^Y_{i'_i} \in \mathcal{GSch}(Y_i, Y_{i'})$, are the given maps).

The maps $\{\varphi_{i_j}\}$ and $\{\varphi'_{i_j}\}$ are considered equivalent if for all $i \in I$, and for $j \geqslant \Sigma(i) \geqslant \sigma(i)$, $\sigma'(i)$ large, $\varphi_{i_j} = \varphi'_{i_j}$. We have functors

$$\begin{aligned} &\varprojlim \; : \mathcal{pro\text{-}BSch} \longrightarrow \mathcal{CBis}_{\mathrm{loc}}/\mathcal{Top} \\ \text{respectively, } &\varprojlim \; : \mathcal{pro\text{-}PSch} \longrightarrow \mathcal{CProp}_{\mathrm{loc}}/\mathcal{Top} \\ &(J, \mathcal{X}_j) \longmapsto \varprojlim_J \mathcal{X}_j \end{aligned} \tag{10.7}$$

The categories $\mathcal{GSch}$ embeds fully faithfully in $\mathcal{pro\text{-}GSch}$: taking $J =$ point.

The categories $\mathcal{pro\text{-}GSch}$ have fiber-products and are closed under limits over directed co-finite posets:

$$\begin{aligned} &(J, \mathcal{X}_j) \prod_{(I,Y_i)} (J', \mathcal{X}'_{j'}) := (J \prod_I J', \mathcal{X}_j \prod_{Y_i} \mathcal{X}'_{j'}) \\ &J \prod_I J' := \{(j, i, j') \in J \prod I \prod J', j \geqslant \sigma(i), j' \geqslant \sigma'(i)\} \end{aligned} \tag{10.8}$$

with $(j_1, i_1, j'_1) \geqslant (j_2, i_2, j'_2)$ iff $i_1 \geqslant i_2$, $j_1 \geqslant j_2$, $j'_1 \geqslant j'_2$.

For limits:

$$\begin{aligned} &\varprojlim_{i \in I} (J_i, \mathcal{X}_j) := \left(\varinjlim_I J_i, \mathcal{X}_j\right) \\ &\varinjlim_I J_i := \{(i, j), i \in I, j \in J_i\} \end{aligned} \tag{10.9}$$

and $(i, j) \geqslant (i', j')$ iff $i \geqslant i'$, $j \geqslant \sigma_{i,i'}(j')$.

Example 10.1. Take

$$J = \{n \in \mathbb{N} \text{ square free}\} = \{p_1 \cdot p_2 \cdots p_\ell,\ p_i \text{ prime}, i \neq j \Rightarrow p_i \neq p_j\} \tag{10.10}$$

with $n_1 \geqslant n_0$ iff n_0 divides n_1. We have the "**compactified** $\mathcal{spec}(\mathbb{Z})$"

$$\overline{\mathcal{spec}(\mathbb{Z})} := \left\{ \mathcal{X}_n = \mathcal{spec}(\mathbb{Z}) \coprod_{\mathcal{spec}(\mathbb{Z}[\frac{1}{n}])} \mathcal{spec}\left(\mathbb{Z}\left[\frac{1}{n}\right] \cap \mathbb{Z}_{\mathbb{R}}\right) \right\}_{n \in J} \tag{10.11}$$

Pictorially, the specialization picture of the underlying topological space $\mathcal{X}_n$ looks like

$\mathcal{X}_n, n = p_1 \cdot p_2 \cdots p_\ell$:

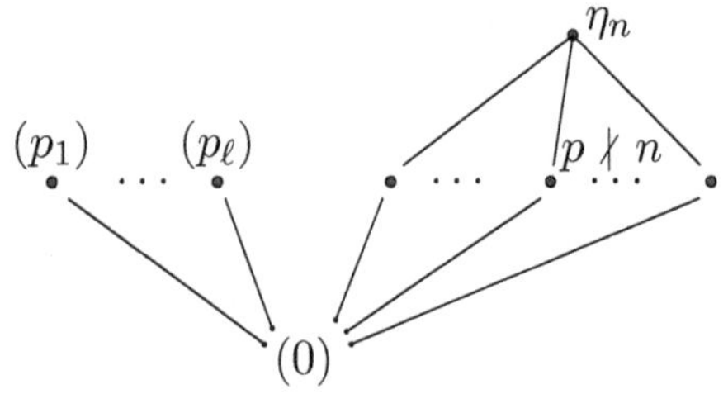

(10.12)

with η_n denoting the maximal ideal of $\mathbb{Z}\left[\frac{1}{n}\right] \cap \mathbb{Z}_{\mathbb{R}}$:

$$\eta_n \equiv \mathbb{Z}\left[\frac{1}{n}\right] \cap (-1, 1).$$

The specialization picuture of the space $\varprojlim_{n} \mathcal{X}_n$ is

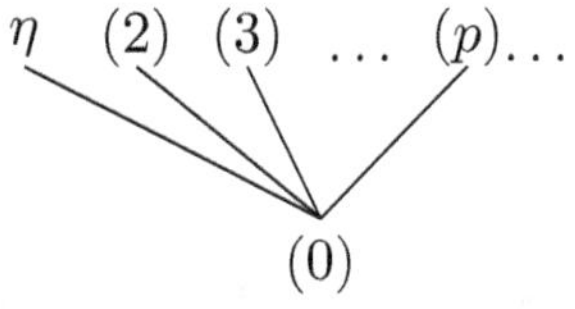

(10.13)

When $n|m$, the map $\mathcal{X}_m \to \mathcal{X}_n$ is the identity on points. Moreover, for $\mathcal{U} \subseteq \mathcal{X}_n$ open: $\mathcal{O}_{\mathcal{X}_n}(\mathcal{U}) \equiv \mathcal{O}_{\mathcal{X}_m}(\mathcal{U})$. But there are more open neighborhood of $\eta_m \in \mathcal{X}_m$ than there are for $\eta_n \in \mathcal{X}_n$.

Clearly $\overline{\mathit{spec}\mathbb{Z}}$ is an element of pro-$\mathcal{P}_T\mathcal{S}ch$, or pro-$\mathcal{B}_T\mathcal{S}ch$, as the sheaves are totally-commutative. The limit sheaf gives for a cofinite open $\mathcal{U} \subseteq \varprojlim \overline{\mathit{spec}(\mathbb{Z})} \equiv \{\eta\} \amalg \mathit{spec}(\mathbb{Z})$ the associated $\mathcal{C}_T\mathcal{P}rop$

$$\mathcal{O}_{\varprojlim \mathit{spec}(\mathbb{Z})}(\mathcal{U})_{n,m} = \left\{a \in \mathbb{Q}_{n,m}\,,\ a \circ \mathbb{Z}_p^m \subseteq \mathbb{Z}_p^n \text{ for all } p \in \mathcal{U}\right\} \quad (10.14)$$

with

$$\mathbb{Z}_\eta^m = \left\{(x_1, \ldots, x_m) \in \mathbb{R}^m,\ \sum_{i=1}^m |x_i|^2 \leqslant 1\right\}$$

the unit ℓ_2-ball for η the real prime.

Similarly, for a number field K, with ring of integers $\mathcal{O}_K$, and with real and complex "primes" $\eta_i : K \hookrightarrow \mathbb{C}$, $i = 1, \ldots, r_{\mathbb{C}} + r_{\mathbb{R}}$

$$\mathbb{R} \bigotimes_{\mathbb{Q}} K \cong \mathbb{C}^{r_{\mathbb{C}}} \times \mathbb{R}^{r_{\mathbb{R}}}, \quad [K : \mathbb{Q}] = 2 \cdot r_{\mathbb{C}} + r_{\mathbb{R}}. \tag{10.15}$$

We take as our partially ordered set the same set J of square free integers.

For $n = p_1 \cdots p_\ell \in J$, let

$$\mathcal{X}_n := \mathit{spec}(\mathcal{O}_K) \coprod_{\mathit{spec}(\mathcal{O}_K[\frac{1}{n}])} \coprod_{i=1}^{r_{\mathbb{C}}+r_{\mathbb{R}}} \mathit{spec}\Big(\mathcal{O}_K\Big[\frac{1}{n}\Big] \cap \eta_i^{-1}(\mathbb{Z}_{\mathbb{C}})\Big) \tag{10.16}$$

We have the "compactified spec $\mathcal{O}_K$",

$$\overline{\mathit{spec}\mathcal{O}_K} \equiv \{\mathcal{X}_n\}_{n \in J} \in \text{pro-}\mathcal{GSch} \tag{10.17}$$

where we can take $\mathcal{G} := \mathcal{B}_T, \mathcal{B}_f, \mathcal{B}, \mathcal{P}_T, \mathcal{P}_f, \mathcal{P}$, as all our bios and props are totally commutative. The topological space

$$\varprojlim_{n \in J} \mathcal{X}_n \equiv \mathit{spec}(\mathcal{O}_K) \amalg \{\eta_i\}_{i=1,\cdots r_{\mathbb{C}}+r_{\mathbb{R}}} \tag{10.18}$$

is one dimensional, the open sets are of the form

$$\mathcal{U} \amalg \{\eta_i\}_{i \in I}, \quad \mathcal{U} \subseteq \mathit{spec}(\mathcal{O}_K) \text{ open}, \quad I \subseteq \{1, \ldots, r_{\mathbb{C}} + r_{\mathbb{R}}\} \tag{10.19}$$

and the sheaf gives

$$\mathcal{O}_{\varprojlim \overline{\mathit{spec}\mathcal{O}_K}}(\mathcal{U} \amalg \{\eta_i\}_{i \in I}) \equiv \mathcal{O}_K(\mathcal{U}) \cap \bigcap_{i \in I} \eta_i^{-1}(\mathbb{Z}_{\mathbb{C}}) \tag{10.20}$$

In particular, the global-sections are

$$\mathcal{O}_{\varprojlim \overline{\mathit{spec}\mathcal{O}_K}}(\varprojlim \overline{\mathit{spec}(\mathcal{O}_K)}) \equiv \mathcal{O}_K \cap \bigcap_{i=1}^{r_{\mathbb{C}}+r_{\mathbb{R}}} \eta_i^{-1}(\mathbb{Z}_{\mathbb{C}}) \equiv \mathbb{F}[\mu_K] \tag{10.21}$$

with $\mu_K \subseteq K^*$ the roots of unity, and as a prop

$$\mathbb{F}[\mu_K]_{n,m} = \left\{ \begin{array}{l} \text{the } n \text{ by } m \text{ matrices with entries in } \{0\} \amalg \mu_K\text{, such} \\ \text{that every row and column contains at most one} \\ \text{non-zero entry} \end{array} \right\} \tag{10.22}$$

Chapter 11

Valuations and Beta Integrals

Definition 11.1. We say $K \in \mathcal{CBio}$ or $\mathcal{CProp}$ is a **field** if

$$K(1)\setminus\{0\} \equiv \mathrm{GL}_1(K) := \left\{a \in K(1), \exists a^{-1} \in K(1), a \circ a^{-1} = 1\right\}$$

and every non-zero scalar is invertible. A sub-Bio or respectively sub-prop $B \subseteq K$ is then a **domain**: $\{0\}$ is prime and the localization gives embeddings

$$B \subseteq B_{(0)} \subseteq K \tag{11.1}$$

We define for such sub-bios

$$\begin{aligned} B^-(n)^\perp &= \{p \in K^-(n), p \circ B^+(n) \subseteq B(1)\} \\ B^+(n)^\perp &= \{q \in K^+(n), B^-(n) \circ q \subseteq B(1)\} \end{aligned} \tag{11.2}$$

and for props

$$B^\perp_{n,m} = \{p \in K_{n,m}, b \circ p \circ d \in B_{1,1} \text{ for all } b \in B_{1,n},\ d \in B_{m,1}\}$$

For fields $k \subseteq K$ we define the **valuation**-sub-bios, respectively, props

$$\mathrm{Val}(K/k) := \left\{ \begin{array}{l} B, \text{(i)}\ \ k \subseteq B \subseteq B_{(0)} \overset{(!)}{\equiv} K \\ \quad \text{(ii)}\ \ B^\pm(n)^\perp = B^\pm(n), \ \ \text{respectively, } B^\perp_{n,m} = B_{n,m} \\ \quad \text{(iii)}\ \ \forall x \in \mathrm{GL}_1(K),\ x \in B(1) \ \text{ or } \ x^{-1} \in B(1) \end{array} \right\} \tag{11.3}$$

We define the **real-valued valuations** for Bios:

$$\mathrm{Val}_{\mathbb{R}}(K/k) := \left\{ \begin{array}{l} |\,| = |\,|_{n,\pm} : K^{\pm}(n) \longrightarrow [0,\infty) \\ |x_1 \circ x_2|_1 = |x_1|_1 \cdot |x_2|_1, \quad |x|_1 = 0 \Leftrightarrow x = 0 \\ |p \circ (p_i)| \leqslant |p| \cdot max\,|p_i|; \quad |(q_i) \circ q| \leqslant |q| \cdot max\,|q_i| \\ |p \overleftarrow{\circ} (q_i)| \leqslant |p| \cdot max\,|q_i|; \quad |(p_i) \overrightarrow{\circ} q| \leqslant |q| \cdot max\,|p_i| \\ \forall \lambda \in k, |\lambda| \leqslant 1 \\ |p|_{n,-} = \sup\{|p \circ q|_1, |q|_{n+} \leqslant 1\} = \inf\{|d^{-1}|_1, |d \circ p|_{n-} \leqslant 1\} \\ |q|_{n,+} = \sup\{|p \circ q|_1, |p|_{n-} \leqslant 1\} = \inf\{|d^{-1}|_1, |q \circ d|_{n+} \leqslant 1\} \end{array} \right\} \Big/ \approx \tag{11.4}$$

and for props:

$$\mathrm{Val}_{\mathbb{R}}(K/k) = \left\{ \begin{array}{l} |\,| = |\,|_{n,m} : K_{n,m} \to [0,\infty) \\ |x_1 \cdot x_2|_{1,1} = |x_1|_{1,1} \cdot |x_2|_{1,1}, \quad |x|_{1,1} = 0 \Leftrightarrow x = 0 \\ |p_1 \circ p_2| \leqslant |p_1| \cdot |p_2| \\ |p_1 \oplus p_2| = max\{|p_1|, |p_2|\} \\ \forall \lambda \in k, \quad |\lambda| \leqslant 1 \\ |p|_{n,m} = \sup\{|b \circ p \circ d|_{1,1}, \quad |b|_{1,n} \leqslant 1, \quad |d|_{m,1} \leqslant 1\} \\ \qquad\quad = \inf\{|d^{-1}|_{1,1}, \quad |d \cdot p|_{n,m} \leqslant 1\} \end{array} \right\} \Big/ \approx$$

where $\approx$ is the equivalence relation given by

$$|| \approx ||^{\sigma}, \quad \sigma > 0$$

Thus in (11.3), condition (i) say that K is the fraction field of the domain B, (ii) say that B is "self-dual", and (iii) is the usual condition that is familiar from rings. Note that in (11.4) the bottom identities, for both bios and props, say that the valuations are completely determined by their (bi)-degree 1 part, and are "self-dual".

We have a map

$$\begin{aligned} \mathrm{Val}_{\mathbb{R}}(K/k) &\hookrightarrow \mathrm{Val}(K/k) \\ |\,| &\longmapsto B_{||} \equiv \{p \in K, |p| \leqslant 1\} \end{aligned} \tag{11.5}$$

This map is a bijection if for all $B \in \mathrm{Val}(K/k)$ the abelian group $\Gamma = \mathrm{GL}_1(K)/\mathrm{GL}_1(B)$, which is totally ordered by

$$x \leqslant y \Longleftrightarrow x \circ y^{-1} \in B(1) \tag{11.6}$$

is "rank one" and there is an order preserving embedding

$$| \ |_{1,1} : \Gamma = \mathrm{GL}_1(K)/\mathrm{GL}_1(B) \hookrightarrow (0, \infty)$$

For any $B \in \mathrm{Val}(K/k)$, we extend the quotient homomorphism to a map

$$\begin{gathered} | \ | : K(1) \longrightarrow K(1)/_{\mathrm{GL}_1(B)} = \{0\} \cup \Gamma \\ |x| = 0 \Longleftrightarrow x = 0 \\ |x_1 \cdot x_2| = |x_1| \cdot |x_2| \\ |1| = 1 = \text{unit of } \Gamma \end{gathered} \tag{11.7}$$

We can embed Γ in a complete ordered abelian group $\Gamma \subseteq \hat{\Gamma}$, so that every subset $\mathscr{A} \subseteq \hat{\Gamma}$ which is bounded above (respectively, below) has a unique least upper bound $\sup \mathscr{A} \in \hat{\Gamma}$ (respectively, maximal lower bound $\inf \mathscr{A} \in \hat{\Gamma}$). In the rank one case we can take $\hat{\Gamma} = (0, \infty)$.

We can than define two maps (we use the notions for Props, the translation to Bios is easy),

$$| \ |_{n,m}, \quad | \ |'_{n,m} : K_{n,m} \longrightarrow \hat{\Gamma} \cup \{0\}$$
$$\text{(respectively, } [0, \infty) \text{ for rank one)}$$

$$|p|_{n,m} = \sup\{|b \circ p \circ d|, \ b \in B_{1,n}, \ d \in B_{m,1}\} \tag{11.8}$$

$$|p|'_{n,m} = \inf\{|s^{-1}|, \ s \in K_{1,1}\backslash\{0\} \ \text{such that} \ s \cdot p \in B_{n,m}\} \tag{11.9}$$

Lemma 11.1. $|p|_{n,m} = |p|'_{n,m}$.

Proof. For $p \in K_{n,m}$, there is $s \in K_{1,1}\backslash\{0\}$ with $s \cdot p \in B_{n,m}$, by the assumption that $B_{(0)} = K$. For $b \in B_{1,n}$, $d \in B_{m,1}$, we have $b \circ (s \cdot p) \circ d \in B_{1,1}$, and so

$$|b \circ p \circ d| = |s^{-1}| \cdot |b \circ (s \cdot p) \circ d| \leqslant |s|^{-1} \tag{11.10}$$

This shown that the set in (11.8) (respectively, (11.9)) is bounded from above (respectively, below), and we can take the sup (respectively, inf). It also shows that we have $|p|_{n,m} \leqslant |p|'_{n,m}$. For the converse inequality, for $p \in K_{n,m}$, any $s \in K_{1,1}\backslash\{0\}$ such that $|s|^{-1} \geqslant |p|_{n,m}$ we have for all

$$b \in B_{1,n}, \ d \in B_{n,1}, \ |s|^{-1} \geqslant |b \circ p \circ d|. \text{ so } b \circ (s \cdot p) \circ d \in B_{1,1},$$

and by our assumption that $B^{\perp} = B$ we get $s \cdot p \in B_{n,m}$, and $|p|'_{n,m} \leqslant |s|^{-1}$ and since s was arbitrary with $|p|_{n,m} \leqslant |s|^{-1}$ we get $|p|'_{n,m} \leqslant |p|_{n,m}$. □

We have

$$|p \circ p'| \leqslant |p| \cdot |p'| \tag{11.11}$$

Indeed, for $s, s' \in K_{1,1} \backslash \{0\}$, such that $|s|^{-1} \geqslant |p|$, $|s'|^{-1} \geqslant |p'|$, we have $s \cdot s' \cdot (p \circ p') = (s \cdot p) \circ (s' \cdot p') \in B$, so $|s|^{-1} \cdot |s'|^{-1} \geqslant |p \circ p'|$.

We have

$$|p_1 \oplus p_2| = max\{|p_1|, |p_2|\} \tag{11.12}$$

Indeed for $s_1, s_2 \in K_{1,1} \backslash \{0\}$, such that $s_i \cdot p_i \in B$, and assume $|s_1| \leqslant |s_2|$, so $s_1 \cdot p_i \in B$, and $s_1 \cdot (p_1 \oplus p_2) = (s_1 \cdot p_1) \oplus (s_1 \cdot p_2) \in B$ and $|p_1 \oplus p_2| \leqslant |s_1|^{-1} = max\{|s_1|^{-1}, |s_2|^{-1}\}$. Taking the infimum over all such s_1, s_2 we get $|p_1 \oplus p_2| \leqslant max\{|p_1|, |p_2|\}$. The inverse inequality follows from (11.11) since we have $p_j = \pi_j \circ (p_1 \oplus p_2) \circ I_j$, with $\pi_j, I_j \in \mathbb{F} \subseteq B$ so

$$|p_j| \leqslant |\pi_j| \cdot |p_1 \oplus p_2| \cdot |I_j| \leqslant |p_1 \oplus p_2|$$

These considerations show that $|\ |$ satisfy all the axioms for a "$\hat{\Gamma} \cup \{0\}$ valued valuation", and in the rank one case $|\,| \in \mathrm{Val}_{\mathbf{R}}(K/k)$.

When the fields $k \subseteq K$ have compatible involution,

$$x \mapsto x^t \text{ for props: } K_{n,m} \xrightarrow{\sim} K_{m,n}, \quad \text{for bios: } K^{\pm}(n) \xrightarrow{\sim} K^{\mp}(n)$$

we let

$$\begin{aligned} &\mathrm{Val}^{\mathbf{t}}\ (K/k) \equiv \{B \in \mathrm{Val}(K/k), B^t \equiv B\} \\ &\mathrm{Val}^{\mathbf{t}}_{\mathbf{R}}(K/k) \equiv \left\{|\,| \in \mathrm{Val}_{\mathbf{R}}(K/k), |x^t| = |x| \text{ all } x \in K\right\} \end{aligned} \tag{11.13}$$

and again, in the rank one situation,

$$\mathrm{Val}^{\mathbf{t}}_{\mathbf{R}}(K/k) \xrightarrow{\sim} \mathrm{Val}^{\mathbf{t}}\ (K/k)$$

Theorem 11.2 (Ostrowski Theorem). *We have*

$$Val^{\mathbf{t}}\ (\mathbb{Q}/\mathbb{F}) = \left\{\mathbb{Q};\ \mathbb{Z}_{(p)} = \mathbb{Q} \cap \mathbb{Z}_p, p\ prime;\ \mathbb{Q} \cap \mathbb{Z}_{\mathbb{R}}\right\} \tag{11.14}$$

For a number field K we have

$$Val^{\mathbf{t}}(K/\mathbb{F}) = \begin{Bmatrix} K;\ \mathcal{O}_{K,\mathfrak{p}} = K \cap \hat{\mathcal{O}}_{K,\mathfrak{p}}, \mathfrak{p} \subseteq \mathcal{O}_K \text{ prime}; \\ K \cap \sigma^{-1}(\mathbb{Z}_{\mathbb{C}}),\ \sigma : K \hookrightarrow \mathbb{C} \quad \text{mod } \sigma \sim \overline{\sigma} \end{Bmatrix} \tag{11.15}$$

Remark 11.1. The requirement that B is preserved by the involution of K, $|p^t| = |p|$, is crucial: without it we have also the ℓ_p-valuations bios $\mathbb{Q} \cap (R_{\ell_q}, \mathbb{R}_{\ell_p})$ or the prop valuations $|\,|_{\ell_p} \in \mathrm{Val}_{\mathbb{R}}(\mathbb{Q}/\mathbb{F})$,

$$|a|_{\ell_p} = \sup\{|a \circ x|_{\ell_p}\,,\ |x|_{\ell_p} \leqslant 1\} \text{ the operator } \ell_p\text{-norm} \tag{11.16}$$

Thus for $\sigma = 1/p \in [0,1]$, we have the "complementary series" of valuation-Props, or Bios, of $\mathbb{Q}$

$$B^{\sigma} = \mathbb{Q} \cap \mathbb{R}_{\ell_{1/\sigma}}, \quad (B^{\sigma})^{\mathrm{op}} = B^{1-\sigma},\ B^{1/2} = \mathbb{Q} \cap \mathbb{Z}_{\mathbb{R}} \tag{11.17}$$

Proof of Ostrowski Theorem. The proof for Bios is a "sub-proof" of the proof for props.

So assume $B \in \mathrm{Val}^t(\mathbb{Q}/\mathbb{F})$ is rank one valuation sub-prop (or Bio) of $\mathbb{Q}$ and let

$$|\,|_{n,m} \ :\ \mathbb{Q}_{n,m} \longrightarrow [0,\infty)$$

be the associated valuation. The "generic point" $B \equiv \mathbb{Q}$, corresponds to the trivial valuation

$$|p|_{n,m} = \begin{cases} 1 & p \neq 0_{n,m} \\ 0 & p = 0_{n,m} \end{cases} \tag{11.18}$$

Assume that $|\,|_{n,m}$ is non-trivial. By our assumption $B = B^t$ we have

$$|(1,1)|_{1,2} = \left|\begin{pmatrix}1\\1\end{pmatrix}\right|_{2,1}$$

Assume first that $|(1,1)|_{1,2} \leqslant 1$. Then for $q_1, q_2 \in \mathbb{Q}$

$$|q_1 + q_2|_{1,1} = \left|(1,1) \circ \begin{pmatrix} q_1 & 0 \\ 0 & q_2 \end{pmatrix} \circ \begin{pmatrix}1\\1\end{pmatrix}\right|_{1,1} \leqslant \mathit{max}\{|q_1|_{1,1}\,,\ |q_2|_{1,1}\}$$

If follows that

$$|n|_{1,1} \leqslant 1, \quad \text{all } n \in \mathbb{Z}$$

Moreover, the set

$$\{n \in \mathbb{Z}, |n|_{1,1} < 1\}$$

is bigger than $\{0\}$ since our valuation is non-trivial, and is a prime ideal $\mathbb{Z} \cdot p \subseteq \mathbb{Z}$, p prime. It follows that

$$\mathbb{Z}_{(p)} \subseteq B_{1,1} \subsetneq \mathbb{Q}$$

But $B_{1,1}$ is closed under addition, and forms an ordinary sub-ring of $\mathbb{Q}$, so

$$\mathbb{Z}_{(p)} = B_{1,1}$$

For any matrix

$$A = (a_{ij}) \in B_{n,m} \subseteq \mathbb{Q}_{n,m}$$

we have

$$a_{ij} = \delta_i \circ A \circ \delta_j^t \in B_{1,1} = \mathbb{Z}_{(p)} \quad \left(\delta_i, \delta_j^t \in \mathbb{F} \subseteq B\right)$$

so

$$B_{n,m} \subseteq \left(\mathbb{Z}_{(p)}\right)_{n,m}, \quad \text{all } n, m$$

Note that

$$|(1,1,\ldots,1)|_{1,n} = \sup\left\{ \left|(1,1,\ldots,1) \circ \begin{pmatrix} a_1 \\ \vdots \\ a_n \end{pmatrix}\right|_{1,1}, \begin{pmatrix} a_1 \\ \vdots \\ a_n \end{pmatrix} \in B_{n,1} \right\}$$

$$\leqslant \sup\left\{|a_1 + a_2 + \cdots + a_n|_{1,1}, a_i \in \mathbb{Z}_{(p)}\right\} \leqslant 1$$

It follows that for any vector $a = (a_1, \ldots, a_n) \in \left(\mathbb{Z}_{(p)}\right)_{1,n}$,

$$|a|_{1,n} = \left|(1,\ldots,1) \circ \begin{pmatrix} a_1 & & 0 \\ & a_2 & \\ & & \ddots \\ 0 & & a_n \end{pmatrix}\right|_{1,n} \leqslant max\{|a_j|_{1,1}\} \leqslant 1$$

and so $\left(\mathbb{Z}_{(p)}\right)_{1,n} \subseteq B_{1,n}$, and we have an equality of Bios $\mathbb{Z}_{(p)} = B$.

For props, take a matrix $A = (a_{i,j}) \in \left(\mathbb{Z}_{(p)}\right)_{n,m}$

$$|A|_{n,m} = \sup\left\{|b \circ A \circ d|_{1,1}\ ,\ b \in \left(\mathbb{Z}_{(p)}\right)_{1,n}\ ,\ d \in \left(\mathbb{Z}_{(p)}\right)_{m,1}\right\} \leqslant 1$$

and so $\left(\mathbb{Z}_{(p)}\right)_{n,m} \subseteq B_{n,m}$ and we have equality of props $B = \mathbb{Z}_{(p)}$.

Assume next $|(1,1)|_{1,2} = \left|\begin{pmatrix}1\\1\end{pmatrix}\right|_{2,1} > 1$.

Passing to an equivalent norm $|\ |^{\lambda}_{1,1}$, we assume $|(1,1)|_{1,2} \leqslant \sqrt{2}$, and

$$|q_1 + q_2|_{1,1} \leqslant 2 \cdot max\{|q_1|_{1,1}, |q_2|_{1,1}\}$$

By induction we get

$$\left|\sum_{i=1}^{2^\gamma} q_i\right|_{1,1} \leqslant 2^\gamma \cdot max\{|q_i|_{1,1}\}$$

so

$$\left|\sum_{i=1}^{n} q_i\right|_{1,1} \leqslant 2 \cdot n \cdot max\{|q_i|_{1,1}\}\ ,\ \text{any } n \geqslant 1$$

It follows that

$$\begin{aligned}
|q_1 + q_2|_{1,1} &= |(q_1+q_2)^n|_{1,1}^{1/n} \\
&= \left|\sum_{k=0}^{n}\binom{n}{k}\cdot q_1^k \cdot q_2^{n-k}\right|_{1,1}^{1/n} \\
&\leqslant (2\cdot(n+1))^{1/n} \cdot max\left(\left|\binom{n}{k}\right|_{1,1} \cdot |q_1|_{1,1}^k \cdot |q_2|_{1,1}^{n-k}\right)^{1/n} \\
&\leqslant (4\cdot(n+1))^{1/n} \cdot max\left(\binom{n}{k} \cdot |q|_{1,1}^k \cdot |q_2|_{1,1}^{n-k}\right)^{1/n} \\
&\leqslant (4\cdot(n+1))^{1/n}\left(|q_1|_{1,1} + |q_2|_{1,1}\right)^{n/n}
\end{aligned}$$

and letting $n \to \infty$ we get the triangle inequality

$$|q_1 + q_2|_{1,1} \leqslant |q_1|_{1,1} + |q_2|_{1,1}$$

We get

$$|(1,1,\ldots,1)|_{1,n} = \sup\left\{ \left|(1,1,\ldots,1) \circ \begin{pmatrix} a_1 \\ \vdots \\ a_n \end{pmatrix}\right|_{1,1}, \left|\begin{pmatrix} a_1 \\ \vdots \\ a_n \end{pmatrix}\right|_{n,1} \leqslant 1 \right\}$$

$$= \sup\left\{ |a_1 + \cdots + a_n|_{1,1}, \left|\begin{pmatrix} a_1 \\ \vdots \\ a_n \end{pmatrix}\right|_{n,1} \leqslant 1 \right\}$$

$$\leqslant \sup\{|a_1|_{1,1} + \cdots + |a_n|_{1,1},\ |a_i|_{1,1} \leqslant 1\} \leqslant n$$

For any $b \in \mathbb{N}$, expanding it in the base $a \in \mathbb{N}$, $a > 1$,

$$b = b_0 + b_1 a + b_2 \cdot a^2 + \cdots + b_m a^m$$

$$0 \leqslant b_j < a, \quad m \leqslant \tfrac{\log b}{\log a}$$

We get

$$|b|_{1,1} = \left|(1,1,\ldots,1) \circ \left(\bigoplus_{j=0}^{m} a^j\right) \circ \left(\bigoplus_{j=0}^{m} b_j\right) \circ (1,1,\ldots,1)^t\right|_{1,1}$$

$$\leqslant |(1,1,\ldots,1)|_{1,m}^2 \cdot \left(\max_{0<d<a} |d|_{1,1}\right) \cdot \max\left\{1, |a|_{1,1}^{\frac{\log b}{\log a}}\right\}$$

$$\leqslant (m+1)^2 \cdot C_a \cdot \max\left\{1, |a|_{1,1}^{\frac{\log b}{\log a}}\right\}$$

with $C_a = \max_{0<d<a} |d|_{1,1}$ a constant depending only on a. It follows that we have

$$|b|_{1,1} = |b^n|_{1,1}^{1/n} \leqslant \left(n \cdot \frac{\log b}{\log a} + 1\right)^{2/n} \cdot C_a^{1/n} \cdot \max\left\{1, |a|_{1,1}^{\frac{\log b}{\log a}}\right\}$$

and letting $n \to \infty$

$$|b|_{1,1} \leqslant max\left\{1, |a|_{1,1}^{\frac{\log b}{\log a}}\right\}$$

If $|b|_{1,1} > 1$ for some $b \in \mathbb{Z}$, then $|a|_{1,1} > 1$ for all $a \in \mathbb{Z}\backslash\{0,1,-1\}$, and we have equality for $a, b \in \mathbb{N}\backslash\{0,1\}$.

$$|b|_{1,1}^{\frac{1}{\log b}} = |a|_{1,1}^{\frac{1}{\log a}} = e^{\delta}$$

so $|b|_{1,1} = |b|^{\delta}$, with $|b|$ the usual real absolute value, and passing to an equivalent norm $|b|_{1,1} = |b|$, any $b \in \mathbb{Z}$, hence for any $b \in \mathbb{Q}$.

We get for vector $q = (q_1, \ldots, q_m) \in \mathbb{Q}_{1,m}$

$$\sum_{j=1}^{m} q_j^2 = |q \circ q^t|_{1,1} \leqslant |q|_{1,m}^2 \quad \text{or} \quad |q|_{1,m} \geqslant \left(\sum_{j=1}^{m} q_j^2\right)^{1/2} = \|q\|$$

where $\|q\|$ is the ℓ_2-norm of q.

On the other hand

$$\begin{aligned} |q|_{1,m} &= \sup\{|q \circ d|_{1,1}\,,\ |d|_{m,1} \leqslant 1\} \\ &\leqslant \sup\left\{\left|\sum_{j=1}^{m} q_j \cdot d_j\right|,\ \sum_{j=1}^{m} d_j^2 \leqslant 1\right\} \\ &= \left(\sum_{j=1}^{m} q_j^2\right)^{1/2} = \|q\| \end{aligned}$$

and we get the equality of Bios $B = \mathbb{Z}_{\mathbb{R}}$.

For props, take a matrix $A = (a_{ij}) \in \mathbb{Q}_{n,m}$, we have

$$\begin{aligned} |A|_{n,m} &= \sup\left\{|b \circ A \circ d|,\ \sum_{i=1}^{n} |b_i|^2 \leqslant 1,\ \sum_{j=1}^{m} |d_j|^2 \leqslant 1\right\} \\ &= \|A\| \text{ the usual operator } \ell_2\text{-norm of } A \end{aligned}$$

and $B = \mathbb{Z}_{\mathbb{R}}$ as props.

For a number field K, let $|| \in \mathrm{Val}^t_{\mathbb{R}}(K/_{\mathbb{F}})$, and let

$$B_{n,m} = \{a \in K_{n,m}, |a|_{n,m} \leqslant 1\}$$

the associated sub-prop (or bio).

For $q_1, q_2 \in K$

$$|q_1 + q_2|_{1,1} = \left|(1,1) \circ \begin{pmatrix} q_1 & 0 \\ 0 & q_2 \end{pmatrix} \circ \begin{pmatrix} 1 \\ 1 \end{pmatrix}\right|_{1,1} \leqslant |(1,1)|^2_{1,2} \cdot max\{|q_1|_{1,1}, |q_2|_{1,1}\}$$

Thus $||_{1,1}$ is a usual valuation of K [**CF67**], hence is either the trivial valuation, or $|q|_{1,1} = |q|_{\mathfrak{p}}$, $\mathfrak{p} \subseteq \mathcal{O}_K$ a finite prime, or $|q|_{1,1} = |\sigma(q)|$ with $\sigma : K \hookrightarrow \mathbb{C}$, modulo conjugation $\sigma \sim \overline{\sigma}$, a "real or complex prime" of K.

In the non-Archimedes case, $|q|_{1,1} = |q|_{\mathfrak{p}}$, $B_{1,1} = \mathcal{O}_{K,\mathfrak{p}}$, and for any matrix $A = (a_{ij}) \in B_{n,m}$

$$|a_{ij}|_{\mathfrak{p}} = |a_{ij}|_{1,1} = |I^t_i \circ A \circ I_j|_{1,1} \leqslant |I^t_i|_{1,n} \cdot |I_j|_{m,1} \cdot |A|_{n,m} \leqslant 1$$

so $B_{n,m} \subseteq (\mathcal{O}_{K,\mathfrak{p}})_{n,m}$.

Note that

$$|(1,\ldots,1)|_{1,n} = \sup\left\{ |a_1 + \cdot + a_n|_{\mathfrak{p}}, \begin{pmatrix} a_1 \\ \vdots \\ a_n \end{pmatrix} \in B_{n,1} \right\}$$

$$\leqslant \sup\{|a_1 + \cdots + a_n|_{\mathfrak{p}}, |a_i|_{\mathfrak{p}} \leqslant 1 \text{ for } i = 1,\ldots,n\} \leqslant 1$$

For a vector $a = (a_1, \ldots, a_n) \in (\mathcal{O}_{K,\mathfrak{p}})_{1,n}$ we get

$$|a|_{1,n} = \left|(1,\ldots,1) \circ \bigoplus_{i=1}^{n} (a_i)\right|_{1,n} \leqslant |(1,\ldots,1)|_{1,n} \cdot max\{|a_i|_{\mathfrak{p}}\} \leqslant 1$$

so

$$(\mathcal{O}_{K,\mathfrak{p}})_{1,n} = B_{1,n}, \text{ and } B = \mathcal{O}_{K,\mathfrak{p}} \text{ as Bios}$$

For a matrix $A = (a_{ij}) \in (\mathcal{O}_{K,\mathfrak{p}})_{n,m}$ we have

$$|A|_{n,m} = \sup\{|b \circ A \circ d|_{\mathfrak{p}},\ b \in (\mathcal{O}_{K,\mathfrak{p}})_{1,n}, d \in (\mathcal{O}_{K,\mathfrak{p}})_{m,1}\} \leqslant 1$$

so $B = \mathcal{O}_{K,\mathfrak{p}}$ also as props.

In the Archimedean case $|q|_{1,1} = |\sigma(q)|$, $\sigma : K \hookrightarrow \mathbb{C}$, we get for a vector $a = (a_1, \ldots, a_n) \in K^n$

$$\sum_{i=1}^{n} |\sigma(a_i)|^2 = |a\bar{a}^t|_{1,1} \leqslant |a|^2_{1,n}$$

so $B_{1,n} \subseteq (\mathcal{O}_{K,\sigma})_{1,n}$ with $\mathcal{O}_{K,\sigma} = K \cap \sigma^{-1}(\mathbb{Z}_{\mathbb{C}})$, and similarly $B_{n,1} \subseteq (\mathcal{O}_{K,\sigma})_{n,1}$. Conversely,

$$\begin{aligned} |a|_{1,n} &= \sup\{|a \circ d|_{1,1}, |d|_{n,1} \leqslant 1\} \\ &\leqslant \sup\left\{\left|\sum_{i=1}^{n} \sigma(a_i \cdot d_i)\right|, \sum_{i=1}^{n} |\sigma(d_i)|^2 \leqslant 1\right\} \equiv \left(\sum_{i=1}^{n} |\sigma(a_i)|^2\right)^{1/2} \end{aligned}$$

so $B_{1,n} = (\mathcal{O}_{K,\sigma})_{1,n}$, and $B = \mathcal{O}_{K,\sigma}$ as Bios.

Finally, for a matrix $A = (a_{ij}) \in K_{n,m}$ we have

$$|A|_{n,m} = \sup\left\{|b \circ A \circ d|, \sum_{i=1}^{n} |\sigma(b_i)|^2 \leqslant 1, \sum_{j=1}^{m} |\sigma(d_j)|^2 \leqslant 1\right\} = \|A\|_{\ell_2}$$

is the usual operator ℓ_2-norm, and $B = \mathcal{O}_{K,\sigma}$ as a Prop too. □

Given a valuation on a number field K, with the associated sub-prop $B_{n,m} = \{x \in K_{n,m}, |x| \leqslant 1\}$, there is a unique metric D on

$$K \cup \{\infty\} \equiv \mathbb{P}^1(K) \equiv \mathbb{P}^1(B) := \{x \in B_{1,2}, |x|_{1,2} = 1\}/_{GL_1(B)} \quad (11.19)$$

which is $GL_2(B)$-invariant and normalized by $D(0, \infty) = 1$.

We obtain a metric on K:

$$d(x, y) = \frac{D(x, y)}{D(x, \infty) \cdot D(y, \infty)} \quad (11.20)$$

(Remembering that K has addition, we have

$$D(x_1, : x_2, y_1 : y_2) = \frac{|x_1 \cdot y_2 - x_2 \cdot y_1|}{|x_1, x_2| \cdot |y_1, y_2|} \quad (11.21)$$

and

$$d(x,y) = |y - x| \tag{11.22}$$

so (11.20) is just a fancy way of not mentioning addition).

We can now normalize the valuation by $|x| = |x|_{1,1} = d(x,0)$.

We can complete K with respect to d, and can assume now that K is complete (every d-Cauchy sequence d-converges). Thus we are passing to the case where K is a **local field**, and its sub-prop of integers B is **compact**: the sets $B_{n,m}$ all have compact topologies, and all the operations are continuous. We have the "spheres"

$$S_K^N \equiv \{x \in B_{1,N}, |x|_{1,N} = 1\} \supseteq GL_N(B) \Big/ GL_N(B)_{(1,0\cdots 0)} \tag{11.23}$$

with

$$GL_N(B)_{(1,0,\cdots 0)} = \{g \in GL_N(B), (1,0,0\cdots 0)g = (1,0,0,\ldots,0)\}$$

the stabilizer.

We say the compact valuation $\mathcal{P}rop\, B$ is **homogeneous** when the action of $\mathrm{GL}_N(B)$ on S_K^N is homogeneous, and we have equality in (11.23). In the number field case, this condition that $\mathrm{GL}_N(B)$ acts transitively on the sphere S_K^N single out at the real and complex places the operator ℓ_2-norm, and $B = \mathbb{Z}_{\mathbb{R}}$ or $\mathbb{Z}_{\mathbb{C}}$, with the usual spheres $S_{\mathbb{R}}^N \equiv S^{N-1}$; $S_{\mathbb{C}}^N \equiv S^{2N-1}$. At the finite primes $\mathfrak{p} \subseteq \mathcal{O}_K$: S_K^N is open-compact in B^N for K non-archimeadian.

The groups $\mathrm{GL}_N(B)$ ($\equiv O(N)$ for $K = \mathbb{R}$; $U(N)$ for $K = \mathbb{C}$; $\mathrm{GL}_N(\mathcal{O}_K)$ for K non-archimedean) are compact, and there is a unique $\mathrm{GL}_N(B)$-invariant probability measure on S_K^N, we denote it by $\sigma_N(dx)$.

For any local field K, with integers the compact homogeneous sub-prop $B \subseteq K$, the spheres $S_K^N \subseteq B_{1,N}$ form a symmetric operad (open: $S_K^o = \varnothing$ empty), sub-operad of B^-,

$$
\begin{aligned}
&o : S_K^n \times S_K^{k_1} \times \cdots \times S_K^{k_n} \longrightarrow S_K^{k_1+\cdots+k_n} \\
&x = (x_1, \ldots, x_n), \quad y_1 = (y_{1,1}, \ldots, y_{1,k_1}), \ldots, \\
&y_n = (y_{n,1}, \ldots, y_{n,k_n}) \rightsquigarrow x \circ (y_i) \\
&\equiv (x_1 \cdot y_{1,1}, \ldots, x_1 \cdot y_{1,k_1}, \ldots, x_i \cdot y_{i,j}, \ldots, x_n \cdot y_{n,1}, \ldots, \; x_n \cdot y_{n,k_n})
\end{aligned} \tag{11.24}
$$

We write the local factors of zeta using the following normalizations

$$
\zeta_K(s) = \begin{cases} 2^{s/2} \cdot \Gamma\left(\frac{s}{2}\right) & K = \mathbb{R} \\ \Gamma(s) & K = \mathbb{C} \\ (1 - q^{-s})^{-1} & K \text{ non-archimedean } q := \#\mathcal{O}_K/\mathfrak{p} \end{cases} \tag{11.25}
$$

We define the multidimensional Beta function

$$
\zeta_K(s_1, \ldots, s_N) := \frac{\zeta_K(s_1) \cdot \zeta_K(s_N)}{\zeta_K(s_1 + \cdots + s_N)} \tag{11.26}
$$

and normalizing it to have the value 1 when $s_1 = \cdots = s_N = 1$, we have

$$
\widetilde{\zeta}_K(s_1, \ldots, s_N) := \frac{\zeta_K(s_1, \ldots, s_N)}{\zeta_K(1, \ldots, 1)} \equiv \frac{\zeta_K(N)}{\zeta_K(1)^N} \cdot \zeta_K(s_1, \ldots, s_N). \tag{11.27}
$$

We have the following basic **Beta-integral**:

$$
\int_{S_K^n} |x_1|^{\alpha_1 - 1} \cdots |x_n|^{\alpha_n - 1} \sigma_n(d(x_1, \ldots, x_n)) \equiv \widetilde{\zeta}_K(\alpha_1, \ldots, \alpha_n) \qquad \operatorname{Re}(\alpha_i) > 0 \tag{11.28}
$$

This can be calculated "by hand" for the different cases of $K = \mathbb{R}$, $K = \mathbb{C}$, and K non-archimedean (there is a q-analogue generalization that gives all these cases as limits, cf. [**Har08**]).

We obtain for $y \in K_{1,n}$.

$$\int_{S_K^n} \left| x \cdot y^t \right|^{\alpha-1} \sigma_n(dx)$$
$$= \int_{S_K^N} |x_1 \cdot y_1 + \cdots + x_n \cdot y_n|^{\alpha-1} \sigma_m\big(d(x, \dots, x_n)\big)$$
$$= \frac{\zeta_K(n)}{\zeta_K(1)} \frac{\zeta_K(\alpha)}{\zeta_K(\alpha+n-1)} |y|^{\alpha-1} \tag{11.29}$$

Indeed, we may assume $y \in S_K^n$, and by $\mathrm{GL}_n(B)$-invariance we may take $y = (1, 0, \dots, 0)$, and (11.29) follows from (11.28) with

$$\alpha_1 = \alpha, \quad \alpha_2 = \cdots = \alpha_n = 1$$

The operad multiplication (11.24) gives the **integration formula**:

$$\int_{S_K^{k_1+\cdots+k_n}} f(x) \sigma_{k_1+\cdots k_n}(dx)$$
$$\equiv \int_{S_K^n} \sigma_n(dx) \frac{|x_1|^{k_1-1} \cdots |x_n|^{k_n-1}}{\widetilde{\zeta}(k_1, \dots, k_n)} \tag{11.30}$$
$$\cdot \int_{S_K^{k_1}} \sigma_{k_1}(dy_1) \cdots \int_{S_K^{k_n}} \sigma_{k_n}(dy_n) f\big(x \circ (y_i)\big)$$

We can generalize both (11.28) and (11.29), writing

$$x = (x_1, \dots, x_n),\ y = (y_1, \dots, y_n) \text{ with } x_i, y_i \in K_{1,k_i},\ k_i \geqslant 1$$

General Beta-integral:

$$\int_{S_K^{k_1+\cdot+k_n}} \left|x_1 \circ y_1^t\right|^{\alpha_1-1} \cdots \left|x_n \circ y_n^t\right|^{\alpha_n-1} \sigma_{k_1+\cdots k_n}(dx)$$
$$= \frac{\zeta_K(k_1+\cdots+k_n)\zeta_K(\alpha_1)\cdots\zeta_K(\alpha_n)}{\zeta_K(1)^n \cdot \zeta_K(\sum(\alpha_i+k_i-1))} |y_1|_{1,k_1}^{\alpha_1-1} \cdots |y_n|_{1,k_n}^{\alpha_n-1} \tag{11.31}$$

(Indeed, this follows from (11.28), (11.29), and (11.30), we omit the straightforward calculation.)

We have the map $\mathbb{N} \longrightarrow K$, and so we have the long vectors $v_n = (\underbrace{1, \dots, 1}_{n})$.

Contracting the measure $\sigma_n(dx)$ on S_K^n by the long vector $v_n^{\mathbf{t}} = (\underbrace{1, 1, \dots, 1}_{n})^{\mathbf{t}}$ we obtain a probability measure on K: for a function $f(x)$ on K,

$$\int_K f(x)\sigma_n \circ v_n^t(dx) :\equiv \int_{S_K^n} f(x \circ v_n^t)\sigma_n(dx)$$

$$\equiv \int_{S_K^n} f(x_1 + \cdots + x_n)\sigma_n(d(x_1, \dots, x_n)) \tag{11.32}$$

We have

$$\sigma_n \circ v_n^t(dx) = \begin{cases} \dfrac{\Gamma\left(\frac{n}{2}\right)}{\sqrt{\pi n}\,\Gamma\left(\frac{n-1}{2}\right)} \cdot \left(1 - \dfrac{|x|^2}{n}\right)_+^{\frac{n-1}{2}-1} dx & K = \mathbb{R},\ dx([0,1]) = 1 \\ \dfrac{n-1}{n\pi} \cdot \left(1 - \dfrac{|x|^2}{n}\right)_+^{n-2} r dr d\theta & K = \mathbb{C},\ x = r \cdot e^{i\theta} \\ \left[\dfrac{1-q^{1-n}}{1-q^{-n}} \mathbf{1}_{\mathcal{O}_K}(x) + \dfrac{q^{1-n}}{1-q^{-n}} \mathbf{1}_{\mathcal{O}_K^*}(x)\right] dx & \begin{array}{c} K \text{ non-archimedean} \\ dx(\mathcal{O}_K) = 1 \end{array} \end{cases} \tag{11.33}$$

With $\mathbf{1}_{\mathcal{O}_K}$ (respectively, $\mathbf{1}_{\mathcal{O}_K^*}$) the characteristic function of $\mathcal{O}_K$ (respectively $\mathcal{O}_K^* = \mathrm{GL}_1(\mathcal{O}_K)$).

Taking the limit $n \longrightarrow \infty$ we obtain the probability measure on K

$$\sigma_\infty(dx) = \lim_{n \longrightarrow \infty} \sigma_n \circ v_n^t(dx),$$

$$\sigma_\infty(dx) = \begin{cases} e^{-\frac{|x|^2}{2}} \dfrac{dx}{\sqrt{2\pi}} & K = \mathbb{R} \\ e^{-|x|^2} \dfrac{r dr d\theta}{\pi} & K = \mathbb{C}, x = r \cdot e^{i\theta} \\ \mathbf{1}_{\mathcal{O}_K}(x) dx & K \text{ non-Archimedean} \end{cases} \tag{11.34}$$

We have from (11.29)

$$\frac{\zeta_K(s)}{\zeta_K(1)} = \int_K |x|^{s-1}\sigma_\infty(dx) = \lim_{n\to\infty}\int_{S_K^n} \left|x \circ v_n^t\right|^{s-1} \sigma_n(dx) \qquad (11.35)$$

Thus for any rank one valuation $||$, on a complete field K, with a compact homogeneous valuation prop $B = \{|x| \leqslant 1\}$ and a map $\mathbb{N} \to K$, (so that we have the vectors v_n), we can **define** the zeta function of B (normalized to have the value 1 for $s = 1$), by

The "Zeta machine":

$$L(B, s) := \lim_{n\to\infty}\int_{S_K^n} |x \circ v_n^t|^{s-1}\sigma_n(dx) \qquad (11.36)$$

with $\sigma_n(dx)$ the unique $\mathrm{GL}_n(B)$-invariant probability measure on the sphere

$$S_K^n = \{x \in B_{1,n}, |x| = 1\}$$

By (11.35) we have $L(\mathcal{O}_K, s) \equiv \zeta_K(s)/\zeta_K(1)$ for $K = \mathbb{R}, \mathbb{C}$ or K non-archimedean.

For a number field K, defining the Dedekind zeta of K by

$$\zeta_K(s) = \prod_{v \in \mathrm{Val}_{\mathbb{R}}(K/\mathbb{F})} \zeta_{K_v}(s) \qquad (11.37)$$

using the normalization (11.25), we have the

$$\textbf{Functional equation: } \zeta_K(1-s) = \left(\frac{D_K}{(2\pi)^{[K:\mathbb{Q}]}}\right)^{\frac{1}{2}-s} \bullet \zeta_K(s) \qquad (11.38)$$

where D_K is the discriminant of K, i.e. $(2\pi)^{[K:\mathbb{Q}]}$ is the contribution to the different of K coming from the real and complex "primes".

For $K = \mathbb{Q}$, the dual of the lattice $\mathbb{Z} \subseteq \mathbb{R}$ with respect to $\langle x, y\rangle \equiv e^{ix\cdot y}$ is the lattice $2\pi \cdot \mathbb{Z}$, and $\log(2\pi)$ is the "genius" of $\overline{spec\mathbb{Z}}$.

The probability measures $\sigma_n \circ v_n^t$ are supported at $\sqrt{n} \cdot \mathbb{Z}_{\mathbb{R}}(1)$, $\sqrt{n}\mathbb{Z}_{\mathbb{C}}(1)$ at the real/complex primes, so they give in the stable-limit

$n \to \infty$ the Gaussian of (11.34). Changing $x \longmapsto \sqrt{n}x$, we make them supported at $\mathbb{Z}_{\mathbb{R}}(1)$, $\mathbb{Z}_{\mathbb{C}}(1)$. Moreover, we can replace the dimension n by the parameter $1+\beta$, and normalize by multiplying by the constant $\frac{\zeta(\beta)}{\zeta(\beta+1)}$, we get the simple expressions for the local field $K_{\mathfrak{p}}$

$$\sigma_\beta(x) = \begin{cases} (1-|x|^2)_+^{\beta/2-1} & K_{\mathfrak{p}} = \mathbb{R} \\ (1-|x|^2)_+^{\beta-1} & K_{\mathfrak{p}} = \mathbb{C} \\ \mathbf{1}_{\mathcal{O}_K}(x) + \frac{q^{-\beta}}{1-q^{-\beta}}\mathbf{1}_{\mathcal{O}_K^*}(x) & K_{\mathfrak{p}} \text{ non-archimedean} \end{cases} \tag{11.39}$$

Now we have the Beta integral written as the Mellin transform of σ_β:

$$\int_{K_{\mathfrak{p}}^*} \sigma_\beta(x)|x|^s\, d^*x \equiv \zeta_{\mathfrak{p}}(\beta, s) \equiv \frac{\zeta_{\mathfrak{p}}(\beta)\zeta_{\mathfrak{p}}(s)}{\zeta_{\mathfrak{p}}(\beta+s)} \tag{11.40}$$

We make σ_β act on functions $\varphi \in \mathcal{S}(K_{\mathfrak{p}})^{\mathcal{O}_K^*}$ via multiplicative convolutions, d^*x multiplicative Haar measure, $d^*x(\mathcal{O}_K^*) = 1$ for non-archimedean K

$$\sigma_\beta\varphi(y) := \int_{K_{\mathfrak{p}}^*} \varphi(x)\sigma_\beta(y/x)|x|^\beta \frac{d^*x}{\zeta_{\mathfrak{p}}(\beta)} \tag{11.41}$$

The Tate distribution $\mathcal{Y}_{\mathfrak{p}}^\beta(dx) \equiv |x|_{\mathfrak{p}}^\beta \frac{d^*x}{\zeta_{\mathfrak{p}}(\beta)}$ satisfies:

$$\mathcal{Y}_{\mathfrak{p}}^s(\sigma_\beta\varphi) \equiv \mathcal{Y}_{\mathfrak{p}}^{s+\beta}(\varphi) \tag{11.42}$$

This shows the semi-group property: $\sigma_{\beta_1}\circ\sigma_{\beta_2}\equiv\sigma_{\beta_1+\beta_2}$, $\lim_{\beta\to 0}\sigma_\beta\equiv\mathrm{Id}$. We have

$$\sigma_\beta(\phi_{\mathfrak{p}}) \equiv \phi_{\mathfrak{p}} \tag{11.43}$$

where the cyclic vector $\phi_{\mathfrak{p}}$, the "vacuum", characterized by $\mathcal{O}_K^*$ invariance, and $\mathcal{Y}_{\mathfrak{p}}^s(\phi_{\mathfrak{p}}) \equiv \mathbf{1}$, is given by the Gaussians cf. (11.34):

$$\phi_{\mathfrak{p}}(z) = \begin{cases} e^{-\frac{x^2}{2}} & K_{\mathfrak{p}} \equiv \mathbb{R} \\ e^{-|x|^2} & K_{\mathfrak{p}} \equiv \mathbb{C} \\ \mathbf{1}_{\mathcal{O}_{K_{\mathfrak{p}}}}(x) & K_{\mathfrak{p}} \text{ non-archimedean} \end{cases}$$

We have similarly a global theory taking:

$$\sigma_\beta(x) \equiv \prod_{\text{all }\mathfrak{p}} \sigma_\beta(x_\mathfrak{p}), \quad x = (x_\mathfrak{p}) \in \mathbb{A}_K$$

Cutting down the action of K^* we get the theory "with intersection":

$$\sigma^!_\beta(x) \equiv \sum_{q \in K^*/\mu_K} \sigma_\beta(qx), \quad x \in K^* \Big\backslash \mathbb{A}^*_K \Big/ \prod_\mathfrak{p} \mathcal{O}^*_{K_\mathfrak{p}} \equiv \mathrm{Pic}_K$$

For example, for $K = \mathbb{Q}$, $\mathrm{Pic}_\mathbb{Q} \equiv \mathbb{Q}^* \Big\backslash \mathbb{A}^*_\mathbb{Q} \Big/ \hat{\mathbb{Z}}^* \cdot \{\pm 1\} \equiv \mathbb{R}^+$,

the global semigroup acts as

$$\sigma^!_\beta \varphi(y) \equiv \int_{\mathbb{R}^+} \varphi(x) \sigma^!_\beta(y/x) |x|^\beta \frac{d^*x}{\zeta_\mathbb{Q}(\beta)}$$

$$\equiv \int_{\mathbb{R}^+} \varphi(x) \left[\sum_{1 \leqslant n \leqslant x/y} \frac{\prod_{p|n}(1-p^\beta)}{2^{\beta/2} \cdot \Gamma(\frac{\beta}{2})} \left(1 - n^2 \frac{y^2}{x^2}\right)_+^{\frac{\beta}{2}-1} \right] |x|^\beta d^*x$$

The Mellin transform

$$\varphi \longmapsto \hat{\varphi}(s) = \int_{\mathbb{R}^+} \varphi(x) |x|^s d^*x \equiv m\varphi(s)$$

gives an isomorphism

$$m: \; L_2(\mathbb{R}^+, d^*x) \xrightarrow{\sim} L^2\left(i\mathbb{R}, \frac{ds}{2\pi i}\right)$$

and we have

$$m\sigma^!_\beta m^{-1} f(s) \equiv \frac{\zeta_\mathbb{Q}(s)}{\zeta_\mathbb{Q}(s+\beta)} f(s+\beta) \equiv \left(\zeta_\mathbb{Q}(s) t_\beta \frac{1}{\zeta_\mathbb{Q}(s)} f\right)(s)$$

with t_β translation by β.

We get for the infinitesimal generator of $\sigma^!_\beta$.

$$m \frac{\partial}{\partial \beta}\Big|_{\beta=0} \sigma^!_\beta m^{-1} f(s) \equiv \frac{\partial}{\partial s} f(s) - f(s) \cdot d \log \zeta_\mathbb{Q}(s)$$

See [**Har08**], Section 6.2.

Chapter 12

Categorical Objects and A-Modules

Let $\mathcal{C}$ be a category with finite products, and final object $*$.

For $A \in \mathcal{P}rop$, we describe explicitly the category $A^{\mathcal{P}}[\mathcal{C}]$ of "**A-prop-objects of $\mathcal{C}$**", it is the subcategory of the functor category $\mathcal{C}^{A\times A^{\mathrm{op}}}$ of objects $\mathfrak{a}$ together with an associative, unital, commutative, "direct-sum" operation

$$\oplus \in \mathcal{C}^{A\times A\times A^{\mathrm{op}}\times A^{\mathrm{op}}}(\mathrm{pr}_{1,3}^{*}\mathfrak{a}\Pi\mathrm{pr}_{2,4}^{*}\mathfrak{a}, (\oplus\times\oplus)^{*}\mathfrak{a})$$

It has objects $\mathfrak{a} = \{\mathfrak{a}_{n,m}\}$, with $\mathfrak{a}_{n,m} \in \mathcal{C}^{S_n\times S_m^{\mathrm{op}}}$, an object of $\mathcal{C}$ with commuting left action of S_n, and right action of S_m, with maps

$$\bigoplus : \mathfrak{a}_{n_1,m_1} \,\Pi\, \mathfrak{a}_{n_2,m_2} \longrightarrow \mathfrak{a}_{n_1+n_2,m_1+m_2} \tag{12.1}$$

which is $S_{n_1}\times S_{n_2} \subseteq S_{n_1+n_2}$, $S_{m_1}\times S_{m_2} \subseteq S_{m_1+m_2}$ covariant, strictly associative, unital $\mathfrak{a}_{0,0} = *$; and $\mathfrak{a}_{n,m} = *$ if $n = 0$ or $m = 0$; and moreover τ-commutative, i.e. we have a commutative diagram

$$\begin{array}{ccc}
\mathfrak{a}_{n_1,m_1}\Pi\mathfrak{a}_{n_2,m_2} & \xrightarrow{\quad\oplus\quad} & \mathfrak{a}_{n_1+n_2,m_1+m_2} \\
\wr\downarrow & & \downarrow \tau_{n_2,n_1}\circ_\circ\tau_{m_1,m_2} \\
\mathfrak{a}_{n_2,m_2}\Pi\mathfrak{a}_{n_1,m_1} & \xrightarrow{\quad\oplus\quad} & \mathfrak{a}_{n_1+n_2,m_1+m_2}
\end{array} \tag{12.2}$$

We have as well A-action, which for $b \in A_{n',n}$, $d \in A_{m,m'}$ associates the map of $\mathcal{C}$,

$$b\circ\_\circ d : \mathfrak{a}_{n,m} \to \mathfrak{a}_{n',m'}$$

this association being $S_n \times S_m^{\mathrm{op}}$-invariant, $S_{n'} \times S_{m'}^{\mathrm{op}}$-covariant, it is associative

$$b_1 \circ (b_2 \circ \underline{\ } \circ d_2)d_1 \equiv (b_1 \circ b_2) \circ \underline{\ } \circ (d_2 \circ d_1) \tag{12.3}$$

unital $\quad 1_n \circ \underline{\ } \circ 1_m \equiv \mathrm{Id}_{\mathfrak{a}_{n,m}}$

(and so we really have commuting left and right A-actions) and functotial: we have commutative diagrams in $\mathcal{C}$.

$$
\begin{array}{ccc}
\mathfrak{a}_{n_1,m_1}\,\Pi\,\mathfrak{a}_{n_2,m_2} & \xrightarrow{(b_1 \circ \underline{\ } \circ d_1)\,\Pi\,(b_2 \circ \underline{\ } \circ d_2)} & \mathfrak{a}_{n_1',m_1'}\,\Pi\,\mathfrak{a}_{n_2',m_2'} \\
{\scriptstyle\oplus}\Big\downarrow & & \Big\downarrow{\scriptstyle\oplus} \\
\mathfrak{a}_{n_1+n_2,m_1+m_2} & \xrightarrow{(b_1 \oplus b_2) \circ \underline{\ } \circ (d_1 \oplus d_2)} & \mathfrak{a}_{n_1'+n_2',m_1'+m_2'}
\end{array}
\tag{12.4}
$$

The arrows $A^P[\mathcal{C}](\mathfrak{a},\mathfrak{a}')$ are given by the natural transformation that preserve the $\oplus$ structures, i.e. by $f = \{f_{n,m}\}$, with

$$f_{n,m} \in \mathcal{C}^{S_n \times S_m^{\mathrm{op}}}(\mathfrak{a}_{n,m}, \mathfrak{a}'_{n,m})$$

arrows of $\mathcal{C}$, commuting with the $S_n \times S_m^{\mathrm{op}}$-actions; strict monoidal:

$$
\begin{array}{ccc}
\mathfrak{a}_{n_1,m_1}\,\Pi\,\mathfrak{a}_{n_2,m_2} & \xrightarrow{f_{n_1,m_1}\,\Pi\,f_{n_2,m_2}} & \mathfrak{a}'_{n_1,m_1}\,\Pi\,\mathfrak{a}'_{n_2,m_2} \\
{\scriptstyle\oplus}\Big\downarrow & & \Big\downarrow{\scriptstyle\oplus} \\
\mathfrak{a}_{n_1+n_2,m_1+m_2} & \xrightarrow{f_{n_1+n_2,m_1+m_2}} & \mathfrak{a}'_{n_1+n_2,m_1+m_2}
\end{array}
\tag{12.5}
$$

is commutative in $\mathcal{C}$; and commutes with the A-action:

$$
\begin{array}{ccc}
\mathfrak{a}_{n,m} & \xrightarrow{f_{n,m}} & \mathfrak{a}'_{n,m} \\
{\scriptstyle b \circ \underline{\ } \circ d}\Big\downarrow & & \Big\downarrow{\scriptstyle b \circ \underline{\ } \circ d} \\
\mathfrak{a}_{n',m'} & \xrightarrow{f_{n'm'}} & \mathfrak{a}'_{n',m'}
\end{array}
\tag{12.6}
$$

Thus, we have a category $A^P[\mathcal{C}]$, the **A-prop-objects of** $\mathcal{C}$.

We will usually assume that $\mathcal{C} \subseteq \mathcal{S}et$, and the product in $\mathcal{C}$ is that of $\mathcal{S}et$, so we can write more easily for the diagrams (12.4), (12.5), (12.6):

$$(b_1 \circ x_1 \circ d_1) \oplus (b_2 \circ x_2 \circ d_2) = (b_1 \oplus b_2) \circ (x_1 \oplus x_2) \circ (d_1 \oplus d_2) \tag{12.4$'$}$$

$$f(x_1 \oplus x_2) = f(x_1) \oplus f(x_2) \tag{12.5$'$}$$

$$f(b \circ x \circ d) = b \circ f(x) \circ d \tag{12.6$'$}$$

For $\mathcal{C} = \mathcal{A}b$, we will write

$$A^P\mathcal{mod} := A^P\left[\mathcal{A}b\right] \tag{12.7}$$

Note that for A^P $\mathcal{mod}$ we have that $b \circ \_ \circ d$ is a homomorphism

$$b \circ (x_1 + x_2) \circ d = (b \circ x_1 \circ d) + (b \circ x_2 \circ d) \tag{12.8}$$

The categories A^P-$\mathcal{mod} = A^P[\mathcal{A}b]$ and $A^P[\mathcal{S}et]$ are complete, and limits are taken pointwise in $\mathcal{S}et$

$$\left(\varprojlim_{j\in J} \mathfrak{a}^{(j)}\right)_{n,m} \equiv \varprojlim_{j\in J}\left(\mathfrak{a}^{(j)}_{n,m}\right) \tag{12.9}$$

Similarly, filtered-co-limits are taken pointwise in $\mathcal{S}et$.

There are also general co-limits.

The category A^P-$\mathcal{mod}$ is abelian.

Remark 12.1. For $M, N, K \in A^P[\mathcal{C}]$, we can define the **Bilinear Maps**

$$\text{Bill}(M, N; K) \equiv \left\{\begin{aligned} &\varphi = \{\varphi^k_{n,m}\},\ \varphi^k_{n,m} \in \mathcal{C}(M_{n,k}\textstyle\prod N_{k,m}, K_{n,m}) \\ &\varphi(m \circ a, n) = \varphi(m, a \circ n) \\ &\varphi(a \circ m, n \circ a') = a \circ \varphi(m, n) \circ a' \\ &\varphi(m_1 \oplus m_2, n_1 \oplus n_2) = \varphi(m_1, n_1) \oplus \varphi(m_2, n_2) \end{aligned}\right\} \tag{12.10}$$

For fix M, N it is a functor of K, and if $\mathcal{C}$ is rich enough, it is representable

$$\mathrm{Bill}(M, N; K) \equiv A^{\mathcal{P}}[\mathcal{C}](M @ N, K) \tag{12.11}$$

$$(M @ N)_{n,m} = \left(\coprod_k M_{n,k} \prod N_{k,m} \right) \Big/ (m \circ a, n) \sim (m, a \circ n) \tag{12.12}$$

This gives a (non-symmetric) monoidal product on $A^{\mathcal{P}}[\mathcal{C}]$.

For example, the associated category of monoids

$$\mathrm{Mon}\left(A^{\mathcal{P}}[\mathcal{S}et_0], @\right) \equiv \mathcal{P}rop_{A\backslash}$$

is precisely the category of props under A, for example,

$$\mathrm{Mon}\left(\mathbb{F}^{\mathcal{P}}[\mathcal{S}et_0], @\right) \equiv \mathcal{P}rop$$

Similarly, for a bio $A = \{A^{\pm}(n)\}$ we have the category $A^{\mathcal{B}}[\mathcal{C}]$ of **A-bio-objects of $\mathcal{C}$**. It has objects $\mathfrak{a} = \{\mathfrak{a}^{\pm}(n)\}$, $\mathfrak{a}^{\pm}(n) \in \mathcal{C}^{S_n}$, object of $\mathcal{C}$ with S_n-action, with left and right multiplication o by A, and left $\overleftarrow{o}$ and right $\overrightarrow{o}$ actions by A (so in total we have maps as in (5.8)) covariant with respect to the S_n-actions, both multiplications and actions are associative and unital, and moreover natural cf. (3.8), (3.9) and co-natural cf. (3.10), (3.11).

Again it follows that we have inverse canonical isomorphism

$$\begin{array}{rcl} \mathfrak{a}^{-}(1) & \xleftrightarrow{\sim} & \mathfrak{a}^{+}(1) \\ x & \longmapsto & x \overrightarrow{o}\, 1^{+} \\ 1^{-} \overleftarrow{o}\, x & \longleftarrow & x \end{array} \tag{12.13}$$

We identify these, write $\mathfrak{a}(1)$ for both, and abuse notations and write $\circ$ for both multiplications and actions.

For objects $\mathfrak{a} = \{\mathfrak{a}^{\pm}(n)\}$, $\mathfrak{a}' = \{\mathfrak{a}'^{\pm}(n)\}$, the arrows $f \in A^{\mathcal{B}}[\mathcal{C}](\mathfrak{a}, \mathfrak{a}')$ are collection $f = \{f^{\pm}(n)\}$, $f^{\pm}(n) \in \mathcal{C}^{S_n}(\mathfrak{a}^{\pm}(n), \mathfrak{a}'^{\pm}(n))$, commuting with all the multiplications and actions of A, so in short

$$f(b \circ x \circ d) \equiv b \circ f(x) \circ d \tag{12.14}$$

Thus we have the category $A^{\mathcal{B}}[\mathcal{C}]$, of **$A$-bio-objects** of $\mathcal{C}$.

When $\mathcal{C} \subseteq \mathcal{S}et$, and the product in $\mathcal{C}$ is that of $\mathcal{S}et$, the A-multiplication and actions are given by a collection of maps as in (5.8). (Indeed, the free $A^{\mathcal{B}}$- $[\mathcal{S}et]$ on one generates of degree 1 is A itself, and the A-bio-subsets of A are the bio-ideals of A.)

Again the categories $A^{\mathcal{B}}[\mathcal{S}et]$ and $A^{\mathcal{B}}[\mathcal{A}b] \equiv A^{\mathcal{B}}\text{-}\mathcal{mod}$ are complete (with limits taken pointwise in $\mathcal{S}et$), and co-complete (with filtered co-limits taken pointwise in $\mathcal{S}et$).

The category $A^{\mathcal{B}}[\mathcal{A}b] \equiv A^{\mathcal{B}}\text{-}\mathcal{mod}$ is abelian.

For $A \in \mathcal{P}rop$ (respectively, $\mathcal{B}io$), $\mathfrak{a} \in A^{\mathcal{P}}[\mathcal{C}]$ (respectively, $A^{\mathcal{B}}[\mathcal{C}]$) is **central** if for $a \in A_{1,1}$ (respectively, $A(1)$), $p \in \mathfrak{a}_{n,m}$, (respectively, $\mathfrak{a}^{\pm}(n)$), we have,

$$a \cdot p := \left(\bigoplus^{n} a\right) \circ p \equiv p \circ \left(\bigoplus^{m} a\right) \tag{12.15}$$

respectively,

$$\begin{aligned} a \cdot p &:= a \circ p \equiv p \circ (\underbrace{a, \ldots, a}_{n}), \quad p \in \mathfrak{a}^{-}(n) \\ a \cdot q &:= q \circ a \equiv (\underbrace{a, \ldots, a}_{n}) \circ q, \quad q \in \mathfrak{a}^{+}(n) \end{aligned} \tag{12.16}$$

In this case, we can localize with respect to a multiplicative subset S of $A_{1,1}$ (respectively, $A(1)$), $(S^{-1}\mathfrak{a})_{n,m} = S^{-1}(\mathfrak{a}_{n,m})$, and we have canonical map

$$\rho_S : \mathfrak{a} \longrightarrow S^{-1}\mathfrak{a} \tag{12.17}$$

which is universal with respect to making multiplication by all $s \in S$ invertible. In particular, we have the localizations

$$\mathfrak{a}_{\mathfrak{p}} := S_{\mathfrak{p}}^{-1}\mathfrak{a}, \quad S_{\mathfrak{p}} = A_{1,1} \backslash \mathfrak{p}, \quad \mathfrak{p} \in \mathcal{s}pec A \tag{12.18}$$

$$\mathfrak{a}\,[1/f] := \{f^{\mathbb{N}}\}^{-1}\mathfrak{a}, \quad f \in A_{1,1} \tag{12.19}$$

For $A \in \mathcal{CP}rop$ (respectively, $\mathcal{CB}io$) we let $\mathcal{C}A^{\mathcal{P}}[\mathcal{C}]$ (respectively, $\mathcal{C}A^{\mathcal{B}}[\mathcal{C}]$) denote the full sub-category of $A^{\mathcal{P}}[\mathcal{C}]$ (respectively, $A^{\mathcal{B}}[\mathcal{C}]$)

of **commutative** objects satisfying for $b \in A_{1,k}$, $d \in A_{k,1}$, $x \in \mathfrak{a}_{n,m}$

$$(b \circ d) \cdot x = \left(\bigoplus^{n} b\right) \circ \sigma_{n,k} \circ \left(\bigoplus^{k} x\right) \circ \sigma_{k,m} \circ \left(\bigoplus^{m} d\right) \tag{12.20}$$

respectively, for $b \in A^-(k)$, $d \in A^+(k)$, $p \in \mathfrak{a}^-(n)$, $q \in \mathfrak{a}^+(n)$

$$\begin{aligned} (b \circ d) \circ p &= b \circ (\underbrace{p, \ldots, p}_{k}) \circ \sigma_{k,n} \circ (\underbrace{d, \ldots, d}_{n}) \\ q \circ (b \circ d) &= (\underbrace{b, \ldots, b}_{n}) \circ \sigma_{n,k} \circ (\underbrace{q, \ldots, q}_{k}) \circ d \end{aligned} \tag{12.21}$$

For a bio $A \in \mathcal{CBio}$, we have the full subcategory $\mathcal{C}_f A^{\mathcal{B}}[\mathcal{C}] \subseteq \mathcal{C}A^{\mathcal{B}}[\mathcal{C}]$ of **fully-commutative** A-object of $\mathcal{C}$, consisting of the $\mathfrak{a} = \{\mathfrak{a}^{\pm}(n)\}$ where we are free to replace any occurrence of $p \circ d$ by

$$\begin{aligned} &(\underbrace{d, \ldots, d}_{n}) \circ \sigma_{n,m} \circ (\underbrace{p, \ldots, p}_{m}), \; p \in \mathfrak{a}^+(n), d \in A^-(m) \\ &\qquad\qquad\text{or} \\ &\qquad\qquad p \in A^+(n),\; d \in \mathfrak{a}^-(m) \end{aligned} \tag{12.22}$$

and similarly replace $b \circ q$ by

$$\begin{aligned} &(\underbrace{q, \ldots, q}_{n}) \circ \sigma_{n,m} \circ (\underbrace{b, \ldots, b}_{m}), \; b \in A^+(n),\; q \in \mathfrak{a}^-(m) \\ &\qquad\qquad\text{or} \\ &\qquad\qquad b \in \mathfrak{a}^+(n),\; q \in A^-(m) \end{aligned}$$

For a prop, $A \in \mathcal{CProp}$ (respectively, $\mathcal{CBio}$) we have the full subcategory $\mathcal{C}_T A^{\mathcal{P}}[\mathcal{C}]$ (respectively, $\mathcal{C}_T A^{\mathcal{B}}[\mathcal{C}] \subseteq \mathcal{C}_f A^{\mathcal{B}}[\mathcal{C}]$) of **totally commutative** A-objects of $\mathcal{C}$, satisfying for $a \in A_{a_0,a_1}$, $x \in \mathfrak{a}_{x_0,x_1}$

$$\sigma_{x_0,a_0} \circ \left(\bigoplus^{a_0} x\right) \circ \sigma_{a_0,x_1} \circ \left(\bigoplus^{x_1} a\right) \equiv \left(\bigoplus^{x_0} a\right) \circ \sigma_{x_0,a_1} \circ \left(\bigoplus^{a_1} x\right) \circ \sigma_{a_1,x_1} \tag{12.23}$$

respectively, we have (12.22) and also

$$\begin{aligned} &p \circ (\underbrace{b, \ldots, b}_{n}) = b \circ (\underbrace{p, \ldots, p}_{m}) \circ \sigma_{m,n}, \; p \in A^-(n), b \in \mathfrak{a}^-(m) \\ &\qquad\qquad\text{or} \\ &\qquad\qquad p \in \mathfrak{a}^-(n),\; b \in A^-(m) \end{aligned}$$

and

$$\begin{aligned} &(\underbrace{d, \ldots, d}_{m}) \circ q = \sigma_{m,n} \circ (\underbrace{q, \ldots, q}_{n}) \circ d, \; q \in A^+(m),\; d \in \mathfrak{a}^+(n) \\ &\qquad\qquad\text{or} \\ &\qquad\qquad q \in \mathfrak{a}^+(m),\; d \in A^+(n) \end{aligned} \tag{12.24}$$

For a prop, we have the category of **fully-commutative** A-object of $\mathcal{C}$, these are the commutative objects that satisfy (12.23) when $a_1 = x_0 = 1$ or $a_0 = x_1 = 1$.

We have the diagram of adjunctions

$$
\begin{array}{ccccccc}
C_T A^P[\mathcal{C}] & \overset{(\;)^T}{\leftrightarrows} & C_f A^P[\mathcal{C}] & \overset{(\;)^f}{\leftrightarrows} & CA^P[\mathcal{C}] & \overset{(\;)^c}{\leftrightarrows} & A^P[\mathcal{C}] \\
\mathcal{F}_T \Big\uparrow \Big\downarrow U & & \mathcal{F}_f \Big\uparrow \Big\downarrow U & & \mathcal{F}_c \Big\uparrow \Big\downarrow U & & \mathcal{F} \Big\uparrow \Big\downarrow U \\
C_T \overline{A}^B[\mathcal{C}] & \leftrightarrows & C_f \overline{A}^B[\mathcal{C}] & \leftrightarrows & C\overline{A}^B[\mathcal{C}] & \leftrightarrows & \overline{A}^B[\mathcal{C}]
\end{array}
$$

the vertical adjunctions when $\overline{A} = UA$ is the bio associated with the prop A.

For a bio A (and hence for a prop $\mathcal{P}$ taking $A = U\mathcal{P}$), we have the simpler notion of an **A-object in $\mathfrak{C}$**, without the prefix "bio" or "prop", and if we want to emphasize we will say "**A-one object of $\mathfrak{C}$**", it is an object of $\mathcal{C}$, $\mathfrak{a} \in \mathcal{C}$, together with A-action:

for $b \in A^-(n)$, $d \in A^+(n)$, we have the maps

$$b \circ \_ \circ d : \underbrace{\mathfrak{a}\pi \cdots \pi\mathfrak{a}}_{n} \longrightarrow \mathfrak{a} \tag{12.25}$$

we will write simply

$$(x_1, \ldots, x_n) \rightsquigarrow b \circ (x_i) \circ d$$

compatible with the S_n-action

$$\mathfrak{b} \circ \sigma \circ (x_{\sigma(i)}) \circ \sigma^{-1} \circ d = b \circ (x_i) \circ d \tag{12.26}$$

associative,

$$(b \circ (b_i)) \circ (x_{i,j}) \circ \big((d_i) \circ d\big) \equiv b \circ (b_i \circ (x_{i,j}) \circ d_i) \circ d \tag{12.27}$$

and unital,

$$1 \circ (x) \circ 1 = x$$

(we leave it to the reader to rewrite these formulas as commutative diagrams for more general categories $\mathcal{C}$).

A map of A-one-objects of $\mathcal{C}$, is a map $f \in \mathcal{C}(\mathfrak{a}, \mathfrak{a}')$ commuting with the A-action

$$f(b \circ (x_i) \circ d) = b \circ \big(f(x_i)\big) \circ d \tag{12.28}$$

or more-generally,

$$\begin{array}{ccc} \mathfrak{a}\pi \cdots \pi \mathfrak{a} & \xrightarrow{\ b \circ \_ \circ d\ } & \mathfrak{a} \\ {\scriptstyle f\pi \cdots \pi f} \downarrow & & \downarrow {\scriptstyle f} \\ \mathfrak{a}'\pi \cdots \pi \mathfrak{a}' & \xrightarrow{\ b \circ \_ \circ d\ } & \mathfrak{a}' \end{array} \tag{12.29}$$

is a commutative diagram in $\mathcal{C}$. Thus, we have the category $A^1[\mathcal{C}]$, of **A-(one)-objects of $\mathcal{C}$.**

The categories $A^1[\mathcal{S}et]$ and $A^1[\mathcal{A}b] \equiv A^1$-$\mathcal{mod}$, are complete and co-complete, and limits and filtered co-limits are taken in $\mathcal{S}et$.

The category A^1-$\mathcal{mod}$ is abelian.

For $A \in \mathcal{CB}i\sigma$, we have the full subcategory of **commutative** A-one-objects of $\mathcal{C}$, $\mathcal{C}A^1[\mathcal{C}] \subseteq A^1\mathcal{C}]$, consisting of the $\mathfrak{a}$'s satisfying for $b \in A^-(n)$, $d \in A^+(n)$,

$$b \circ (x, \ldots, x) \circ d = b \circ d \circ (x) \circ 1 = 1 \circ (x) \circ b \circ d \tag{12.30}$$

We have as well the full subcategory of **fully-commutative** one objects $\mathcal{C}_f A^1[\mathcal{C}]$ satisfying for $d \in A^+(k)$, $d_i \in A^+(m_i)$, $b \in A^-(m_1 + \cdots + m_k)$,

$$b \circ \left(\underbrace{x_1, \ldots, x_1}_{m_1}, \underbrace{x_2, \ldots, x_2}_{m_2}, \ldots, \underbrace{x_k, \ldots, x_k}_{m_k} \right) \circ \big((d_1, \ldots, d_k) \circ d\big)$$
$$\equiv \big(b \overleftarrow{\circ} (d_i)\big) \circ (x_i) \circ d \tag{12.31}$$

and similarly, for $b \in A^-(k)$, $b_i \in A^-(m_i)$, $d \in A^+(m_1 + \cdots + m_k)$,

$$(b \circ (b_1, \ldots, b_k)) \circ \left(\underbrace{x_1, \ldots, x_1}_{m_1}, \underbrace{x_2, \ldots, x_2}_{m_2}, \ldots, \underbrace{x_k, \ldots, x_k}_{m_k} \right) \circ d$$
$$\equiv b \circ (x_i) \circ \big((b_i) \overrightarrow{\circ} d\big) \tag{12.32}$$

A fully commutative A-one-object of $\mathcal{C}$ will be called **totally-commutative** if for $b \in A^-(n)$, $d \in A^+(n)$, $p \in A^-(m)$, $q \in A^+(m)$, and elements x_k, $k = 1, \ldots, n \cdot m$, we have

$$\begin{aligned} & b \circ (\underbrace{p, \ldots, p}_{n}) \circ (x_k) \circ (\underbrace{q \cdots q}_{n}) \circ d \\ & = p \circ (\underbrace{b, \ldots, b}_{m}) \circ (x_{\sigma_{m,n}(k)}) \circ (\underbrace{d, \ldots, d}_{m}) \circ q \end{aligned} \tag{12.33}$$

Note the permutation $\sigma_{n,m}$ of (0.27) appearing in the indices of the x_k's. This last condition is automatic when A is totally-commutative.

We have the full embeddings and their left adjoints

$$\mathcal{C}_T A^1[\mathcal{C}] \underset{()^T}{\leftrightarrows} \mathcal{C}_f A^1[\mathcal{C}] \underset{()^f}{\leftrightarrows} \mathcal{C}A^1[\mathcal{C}] \underset{()^C}{\leftrightarrows} A^1[\mathcal{C}]$$

Given a homomorphism $\varphi \in \mathcal{P}rop(B, A)$, respectively, $\mathcal{B}io(B, A)$, we get

$$\begin{aligned} & B^P[\mathcal{C}] \xleftarrow{\varphi^*} A^P[\mathcal{C}], \\ & \text{respectively, } B^B[\mathcal{C}] \xleftarrow{\varphi^*} A^B[\mathcal{C}] \quad \text{or} \quad B^1[\mathcal{C}] \xleftarrow{\varphi^*} A^1[\mathcal{C}] \end{aligned} \tag{12.34}$$

via pull-back of structure,

$$b_1 \circ (x_i) \circ b_2 := \varphi(b_1) \circ (x_i) \circ \varphi(b_2), \; b_i \in B \tag{12.35}$$

The pull-back functor φ^* preserves the various commutativity sub-categories.

When $\mathcal{C}$ has colimits, and in particular $\mathcal{C} = \mathcal{S}et$ or $\mathcal{C} = \mathcal{A}b$, these functors will have Kan left adjoins, the "extension of scalars", e.g. for props,

$$\begin{array}{ccccc} \mathcal{C}_T A^P[\mathcal{C}] & \underset{()^T}{\leftrightarrows} & \mathcal{C}A^P[\mathcal{C}] & \underset{()^C}{\leftrightarrows} & A^P[\mathcal{C}] \\ {\scriptstyle A \underset{B}{\boxtimes}^T -} \uparrow \downarrow {\scriptstyle \varphi^*} & & {\scriptstyle A \underset{B}{\boxtimes}^C -} \uparrow \downarrow {\scriptstyle \varphi^*} & & {\scriptstyle A \underset{B}{\boxtimes} -} \uparrow \downarrow {\scriptstyle \varphi^*} \\ \mathcal{C}_T B^P[\mathcal{C}] & \underset{()^T}{\leftrightarrows} & \mathcal{C}B^P[\mathcal{C}] & \underset{()^C}{\leftrightarrows} & B^P[\mathcal{C}] \end{array} \tag{12.36}$$

Given $(\mathfrak{a}, \mathfrak{b}) \in A^P[\mathcal{C}] \times B^P[\mathcal{C}]$, respectively, $A^B[\mathcal{C}] \times B^B[\mathcal{C}]$ or $A^1[\mathcal{C}] \times B^1[\mathcal{C}]$, the map

$$f = \{f_{n,m}\}, f_{n,m} \in \mathcal{C}^{S_n \times S_m^{\text{op}}}(\mathfrak{b}_{n,m}, \mathfrak{a}_{n,m})$$

respectively,

$$f_n^{\pm} \in \mathcal{C}^{S_n}\big(\mathfrak{b}^{\pm}(n), \mathfrak{a}^{\pm}(n)\big) \text{ or } f \in \mathcal{C}(\mathfrak{b}, \mathfrak{a})$$

is said to be **compatible** with $\varphi : B \to A$, if $f \in B^P[\mathcal{C}](\mathfrak{b}, \varphi^*\mathfrak{a})$, respectively, $B^B[\mathcal{C}](\mathfrak{b}, \varphi^*\mathfrak{a})$ or $B^1[\mathcal{C}](\mathfrak{b}, \varphi^*\mathfrak{a})$.

Given a prop space $(X, \mathcal{O}_X) \in \mathcal{CProp}/\mathcal{Top}$ (respectively, bio space in $\mathcal{CBio}/\mathcal{Top}$) we have the categories of $\mathcal{O}_X$-prop (respectively, bio or one) — objects of $\mathcal{C}$, $(\mathcal{O}_X)^P[\mathcal{C}]$, (respectively, $(\mathcal{O}_X)^B[\mathcal{C}]$ or $(\mathcal{O}_X)^1[\mathcal{C}]$) it has objects $\mathfrak{a} = \mathfrak{a}(\mathcal{U})$ a functor associating with an open set $\mathcal{U} \subseteq X$ the object $\mathfrak{a}(\mathcal{U}) \in \mathcal{O}_X(\mathcal{U})^P[\mathcal{C}]$ (respectively, $\mathcal{O}_X(\mathcal{U})^B[\mathcal{C}]$ or $\mathcal{O}_X(\mathcal{U})^1[\mathcal{C}]$) and for $\mathcal{V} \subseteq \mathcal{U}$ we have restriction maps $r_{\mathcal{V}}^{\mathcal{U}} \in \mathcal{O}_X(\mathcal{U})[\mathcal{C}]\Big(\mathfrak{a}(\mathcal{U}), (\rho_{\mathcal{V}}^{\mathcal{U}})^*\mathfrak{a}(\mathcal{V})\Big)$, compatible with the restrictions $\rho_{\mathcal{V}}^{\mathcal{U}} : \mathcal{O}_X(\mathcal{U}) \to \mathcal{O}_X(\mathcal{V})$, satisfying functoriality

$$r_{\mathcal{W}}^{\mathcal{V}} \circ r_{\mathcal{V}}^{\mathcal{U}} = r_{\mathcal{W}}^{\mathcal{U}} \text{ for } \mathcal{W} \subseteq \mathcal{V} \subseteq \mathcal{U} \ ; \ r_{\mathcal{U}}^{\mathcal{U}} = \text{id} \tag{12.37}$$

and $\mathcal{U} \mapsto \mathfrak{a}(\mathcal{U})$ forming a **sheaf** (which, when $\mathcal{C} \subseteq \mathcal{Set}$, with products in $\mathcal{C}$ created by $\mathcal{Set}$, means that for all n, m, $\mathcal{U} \mapsto \mathfrak{a}_{n,m}(\mathcal{U})$ forms a sheaf of $\mathcal{Set}$ over X).

Assume now $\mathcal{C}$ has filtered co-limits. For an object $\mathfrak{a} \in \mathcal{C}$, and (a self map) $s \in \mathcal{C}(\mathfrak{a}, \mathfrak{a})$, we have the localization map

$$\mathfrak{a} \longrightarrow \mathfrak{a}\left[\frac{1}{s}\right] := \lim_{\to}\left\{\mathfrak{a} \xrightarrow{\cdot s} \mathfrak{a} \xrightarrow{\cdot s} \mathfrak{a} \xrightarrow{\cdot s} \cdots\right\} \tag{12.38}$$

For a commutative monoid S, with an action on $\mathfrak{a}$, i.e. we have a map of monoids $S \to \mathcal{C}(\mathfrak{a}, \mathfrak{a})$, we have the localization map

$$\mathfrak{a} \longrightarrow S^{-1}\mathfrak{a} := \lim_{\substack{\longrightarrow \\ s \in S}} \mathfrak{a}\left[\frac{1}{s}\right] \tag{12.39}$$

For $A \in \mathcal{CProp}$, respectively, $\mathcal{CBio}$, and a multiplicative set $S \subseteq A_{1,1}$, respectively, $A(1)$, we get localization maps, compatible with the localization homomorphism $\varrho_S : A \to S^{-1}A$, and giving the extension of scalars

$$(\varrho_S)_* \equiv S^{-1}A \boxtimes_A^{\mathcal{C}} \_ (= \text{the left adjoint of } \varrho_S^*) : \mathcal{C}A^{\mathcal{P}}[\mathbb{C}] \to \mathcal{C}(S^{-1}A)^{\mathcal{P}}[\mathbb{C}],$$
$$\text{respectively, } \mathcal{C}A^{\mathcal{B}}[\mathbb{C}] \to \mathcal{C}(S^{-1}A)^{\mathcal{B}}[\mathbb{C}] \text{ or } \mathcal{C}A^{1}(\mathbb{C}) \to \mathcal{C}(S^{-1}A)^{1}(\mathbb{C}),$$
$$\mathfrak{a}_{n,m} \mapsto S^{-1}\mathfrak{a}_{n,m} \qquad \mathfrak{a}^{\pm}(n) \mapsto S^{-1}\mathfrak{a}^{\pm}(n), \text{ or just } \mathfrak{a} \mapsto S^{-1}\mathfrak{a} \tag{12.40}$$

These S-localization functors commute with any co-limits (being left adjoints), and also commute with finite limits (since we can take "common denominators").

We will assume from now on that $\mathbb{C} \subseteq \mathcal{S}et$, limits in $\mathbb{C}$ are created in $\mathcal{S}et$, and filtered co-limits are created in $\mathcal{S}et$.

Given $A \in \mathcal{CP}rop$, (respectively, $\mathcal{CB}io$), and $\mathfrak{a} \in \mathcal{C}A^{\mathcal{P}}[\mathbb{C}]$, (respectively, $\mathcal{C}A^{\mathcal{B}}[\mathbb{C}]$ or$\mathcal{C}A^{1}[\mathbb{C}]$) we get $\tilde{\mathfrak{a}} \in \mathcal{C}(\mathcal{O}_A)^{\mathcal{P}}[\mathbb{C}]$, (respectively, $\mathcal{C}(\mathcal{O}_A)^{\mathcal{B}}[\mathbb{C}]$ or $\mathcal{C}(\mathcal{O}_A)^{1}(\mathbb{C})$), by defining for $\mathcal{U} \subseteq \mathit{spec}A$ open

$$\tilde{\mathfrak{a}}(\mathcal{U})_{n,m} = \left\{ f : \mathcal{U} \to \coprod_{\mathfrak{p} \in \mathcal{U}} S_{\mathfrak{p}}^{-1}\mathfrak{a}_{n,m},\ f(\mathfrak{p}) \in S_{\mathfrak{p}}^{-1}\mathfrak{a}_{n,m} \text{ locally a fraction} \right\} \tag{12.41}$$

***f* locally a fraction**: for all $\mathfrak{p} \in \mathcal{U}$, there exists open $\mathfrak{p} \in \mathcal{U}_{\mathfrak{p}} \subseteq \mathcal{U}$, $a \in \mathfrak{a}_{n,m}$, $s \in A_{1,1} \setminus \bigcup_{\mathfrak{q} \in \mathcal{U}_{\mathfrak{p}}} \mathfrak{q}$, such that for all $\mathfrak{q} \in \mathcal{U}_{\mathfrak{p}}$:

$$f(\mathfrak{q}) \equiv a/s \in S_{\mathfrak{q}}^{-1}\mathfrak{a}_{n,m}$$

There is a natural action of $\mathcal{O}_A(\mathcal{U})$ on $\tilde{\mathfrak{a}}(\mathcal{U})$, the "locally-fraction" condition is preserved; there are restriction maps $\tilde{\mathfrak{a}}(\mathcal{U}) \to \tilde{\mathfrak{a}}(\mathcal{V})$ for $\mathcal{V} \subseteq \mathcal{U}$ compatible with $\mathcal{O}_A(\mathcal{U}) \to \mathcal{O}_A(\mathcal{V})$; and clearly $\mathcal{U} \to \tilde{\mathfrak{a}}(\mathcal{U})$ is a sheaf.

Theorem 12.1. *For $A \in \mathcal{CP}rop$ (respectively, $\mathcal{CB}io$), $\mathfrak{a} \in \mathcal{C}A^{\mathcal{P}}[\mathbb{C}]$ (respectively, $\mathcal{C}A^{\mathcal{B}}[\mathbb{C}]$ or $\mathcal{C}A^{1}[\mathbb{C}]$), with $\mathbb{C} \subseteq \mathcal{S}et$ having limits and filtered co-limits created by $\mathcal{S}et$, there exists a sheaf $\tilde{\mathfrak{a}}$ over $\mathit{spec}A$, with $\tilde{\mathfrak{a}}(\mathcal{U}) \in \mathcal{O}_A(\mathcal{U})[\mathbb{C}]$, with the $\mathcal{O}_A(\mathcal{U})$-structure compatible with respect to restrictions $\tilde{\mathfrak{a}}(\mathcal{U}) \to \tilde{\mathfrak{a}}(\mathcal{V})$, $\mathcal{O}_A(\mathcal{U}) \to \mathcal{O}_A(\mathcal{V})$, $\mathcal{V} \subseteq \mathcal{U} \subseteq \mathit{spec}A$*

open, with stalks

$$\tilde{\mathfrak{a}}\Big|_{\mathfrak{p}} := \underset{\mathfrak{p}\in\mathcal{U}\subseteq \mathit{spec}A}{colim} \tilde{\mathfrak{a}}(\mathcal{U}) \xrightarrow{\sim} \mathfrak{a}_{\mathfrak{p}} \equiv S_{\mathfrak{p}}^{-1}\mathfrak{a} \tag{12.42}$$

and global sections

$$\mathfrak{a}(D_A(f)) \xleftarrow{\sim} \mathfrak{a}[1/f] \tag{12.43}$$

Proof. Exactly as in the proof of Theorem 8.1.

All we need is commutativity! □

Theorem 12.2. *We have adjunction*

$$\begin{array}{ccc} \tilde{\mathfrak{a}} & \mathcal{C}\mathcal{O}_A[\mathcal{C}] & \mathcal{O} \\ \big\uparrow & \uparrow\;\downarrow & \big\downarrow \\ \mathfrak{a} & \mathcal{C}A[\mathcal{C}] & \mathcal{O}(\mathit{spec}A) =: \Gamma\mathcal{O} \end{array}$$

Proof. Similar to the proof of Theorem 9.1. □

For a generalized scheme $(\mathcal{X}, \mathcal{O}_{\mathcal{X}}) \in \mathcal{PSch}$, respectively, $\mathcal{BSch}$, and $\mathfrak{a} \in (\mathcal{O}_{\mathcal{X}})^{P}[\mathcal{C}]$, respectively, $(\mathcal{O}_{\mathcal{X}})^{B}[\mathcal{C}]$ or $(\mathcal{O}_{\mathcal{X}})^{1}[\mathcal{C}]$, the following conditions are equivalent:

(i) For some open affine covering $\mathcal{X} = \bigcup_i \mathit{spec}A_i$, we have

$$\mathfrak{a}|_{\mathit{spec}A_i} \equiv \mathfrak{a}(\mathit{spec}A_i)^{\sim}$$

(ii) For any open affine $\mathit{spec}A \subseteq \mathcal{X}$ we have (12.44)

$$\mathfrak{a}|_{\mathit{spec}A} \equiv \mathfrak{a}(\mathit{spec}A)^{\sim}$$

(iii) For any open $\mathcal{U} \subseteq \mathcal{X}$, and any section $s \in \mathcal{O}_{\mathcal{X}}(\mathcal{U})_{1,1}$, let

$$D(s) := \{x \in \mathcal{U},\ s|_x \notin \mathfrak{m}_x\}$$

denote the **open** subset where s is invertible, then restriction from $\mathcal{U}$ to $D(s)$ induces isomorphism

$$\mathfrak{a}(\mathcal{U})[1/s] \xrightarrow{\sim} \mathfrak{a}\big(D(s)\big)$$

We say $\mathfrak{a}$ is **quasi-coherent** if it satisfies these conditions, and we have the full-subcategories of quasi-coherent objects

$$q.c.(\mathcal{O}_X)^P[\mathcal{C}] \subseteq \mathcal{C}(\mathcal{O}_X)^P[\mathcal{C}], \quad q.c.(\mathcal{O}_X)^B[\mathcal{C}] \subseteq \mathcal{C}(\mathcal{O}_X)^B[\mathcal{C}],$$
$$q.c.(\mathcal{O}_X)^1[\mathcal{C}] \subseteq \mathcal{C}(\mathcal{O}_X)^1[\mathcal{C}]$$

For $\mathcal{X} = \mathit{spec} A$ affine, we have equivalence

$$\mathcal{C}A^P[\mathcal{C}] \xleftrightarrow{\sim} q.c.(\mathcal{O}_A)^P[\mathcal{C}] \quad \text{respectively,}$$
$$\mathcal{C}A^B[\mathcal{C}] \xleftrightarrow{\sim} q.c.(\mathcal{O}_A)^B[\mathcal{C}] \tag{12.45}$$

Examples 12.1 (of one-objects).
$*$ For any commutative bio A, $A(1)$ is the free A-set on one generator, and its $A^1[\mathcal{S}et]$-subsets are the ideals of A.
$*$ We have $\mathbb{F}^1[\mathcal{S}et] \equiv \mathcal{S}et_0$, the category of pointed-sets, and every pointed set, $0_X \in X$, has a unique structure as $\mathbb{F}$-one-set:

$$\delta_i \circ (x_k) \circ \delta_j^t = \begin{cases} x_i & i = j \\ 0_X & i \neq j \end{cases} \tag{12.46}$$

$*$ For $A \in \mathcal{CR}ing$, we have $A^1[\mathcal{S}et_0] \equiv A^1[\mathcal{A}b] \equiv A\text{-}\mathcal{mod}$, the usual A-modules, and every A-module M gives elements of $A^1[\mathcal{A}b]$, respectively, $A^B[\mathcal{A}b]$, $A^P(\mathcal{A}b)$, by taking M, respectively,

$$M^{\pm}(n) := \text{row/column vectors with values in } M \tag{12.47}$$
$$M_{n,m} := n \text{ by } m \text{ matrices with values in } M \tag{12.48}$$

These are always totally-commutative objects of the prop/bio/one A-object of $\mathcal{A}b$, for A commutative.

$*$ The $\mathbb{Z}_{\mathbb{R}}$ (respectively, $\mathbb{Z}_{\mathbb{C}}$) modules which are "torsion free" and embedded as sub-sets of a real (respectively, complex) vector space $V \in \mathbb{R}^1[\mathcal{Ab}] \equiv \mathbb{R}^1[\mathcal{Set}_0]$, (respectively, $\mathbb{C}^1[\mathcal{Ab}]$), are the **convex symmetric subsets** $C \subseteq V$,

$$\begin{aligned} &c_1 \cdots c_n \in C, \;\; t_i \geqslant 0, \;\; \sum_{i=1}^{n} t_i = 1 \Longrightarrow \sum_{i=1}^{n} t_i \cdot c_i \in C \\ &u \cdot C = C \text{ for } u = \pm 1 \;\; (\text{respectively, } |u| = 1) \end{aligned} \tag{12.49}$$

This follows from $b \circ (c_i) \circ d \in C$, for any b, d in the unit ℓ_2-ball, noting that the pointwise product $(b_1 \cdot d_1, \ldots, b_n \cdot d_n)$ is an arbitrary vector in the unit ℓ_1-ball.

Example 12.2. For a map of generalized schemes $f : X \to Y$, the functor f^* takes quasi-coherent $\mathcal{O}_Y$-objects of $\mathcal{C}$ to quasi-coherent $\mathcal{O}_X$-objects of $\mathcal{C}$, and moreover, for affine open subsets $\mathcal{U} = \mathit{spec} A \subseteq X$, $\mathcal{V} = \mathit{spec} B \subseteq Y$, with $f(\mathcal{U}) \subseteq \mathcal{V}$, and with $\varphi = f^{\#} : B \to A$ the associated homomorphism, we have for quasi-coherent sheaf M of objects of $\mathcal{C}$, on Y

$$f^* M \Big|_{\mathcal{U}} \cong (\varphi_* M)^{\sim} (\mathcal{V})$$

When the map f is quasi-coherent, i.e. quasi-compact and quasi-separated cf. (15.10, vii), the left adjoin functor f_* of f^* also takes quasi-coherent $\mathcal{O}_X$-objects of $\mathcal{C}$ to quasi coherent $\mathcal{O}_Y$-objects of $\mathcal{C}$, cf., Remark 15.2.

Chapter 13

Derivations and Differential

Given $A \in \mathcal{P}rop$, $M \in A^P\text{-}mod$ (respectively, $A \in \mathcal{B}io$, $M \in A^B\text{-}mod$), we can form the product $A \sqcap M$ pointwise

$$(A \sqcap M)_{n,m} = A_{n,m} \sqcap M_{n,m}, \quad \text{respectively,}$$
$$(A \sqcap M)^{\pm}(n) = A^{\pm}(n) \sqcap M^{\pm}(n) \tag{13.1}$$

It has the structure of a prop (bio) over A

$$(a_1, m_1) \circ (a_2, m_2) = (a_1 \circ a_2, m_1 \circ a_2 + a_1 \circ m_2)$$
$$(a_1, m_1) \oplus (a_2, m_2) = (a_1 \oplus a_2, m_1 \oplus m_2) \tag{13.2}$$

$$\pi : A \sqcap M \longrightarrow\!\!\!\!\!\rightarrow A, \quad \pi(a, m) = a$$

Moreover, it is an Abelian group object of the category $\mathcal{P}rop/A$ respectively, $\mathcal{B}io/A$, via

addition: $\mu_+ : (A \sqcap M) \prod_A (A \sqcap M) \equiv A \sqcap M \sqcap M \longrightarrow A \sqcap M$

$$(a, m_1), (a, m_2) \longleftrightarrow (a, m_1, m_2) \rightsquigarrow (a, m_1 + m_2) \tag{13.3}$$

zero ε**:** $\varepsilon(a) = (a, 0)$

$$\begin{array}{ccc} & A \sqcap M & \\ \varepsilon \nearrow & & \searrow \pi \\ A & \xrightarrow{\quad \mathrm{id}_A \quad} & A \end{array} \tag{13.4}$$

inverse: $S : A \sqcap M \longrightarrow A \sqcap M$, $S(a, m) = (a, -m)$ \tag{13.5}

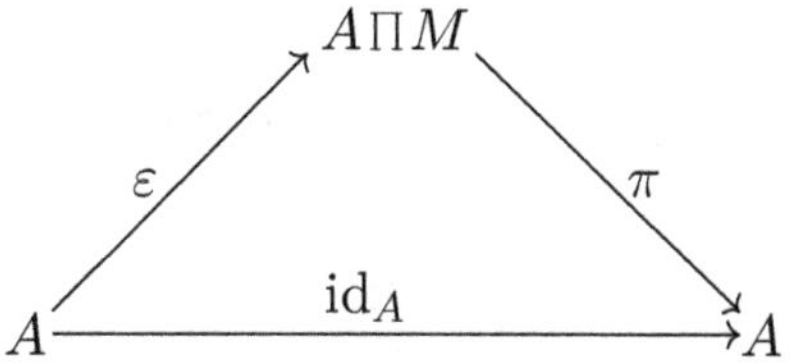

The commutativity of $A \Pi M$ can be analyzed, and we have

$$\begin{array}{lll} P(\text{i}) & A \in \mathcal{CProp}, & M \in \mathcal{C}_T A^P\text{-}\mathcal{mod} \Longrightarrow A\Pi M \in \mathcal{CProp} \\ P(\text{ii}) & A \in \mathcal{C}_f\mathcal{Prop}, & M \in \mathcal{C}_T A^P\text{-}\mathcal{mod} \Longrightarrow A\Pi M \in \mathcal{C}_f\mathcal{Prop} \\ P(\text{iii}) & A \in \mathcal{C}_T\mathcal{Prop}, & M \in \mathcal{C}_T A^P\text{-}\mathcal{mod} \Longrightarrow A\Pi M \in \mathcal{C}_T\mathcal{Prop} \\ B(\text{i}) & A \in \mathcal{CBio}, & M \in \mathcal{C}_f A^B\text{-}\mathcal{mod} \Longrightarrow A\Pi M \in \mathcal{CBio} \\ B(\text{ii}) & A \in \mathcal{C}_f\mathcal{Bio}, & M \in \mathcal{C}_f A^B\text{-}\mathcal{mod} \Longrightarrow A\Pi M \in \mathcal{C}_f\mathcal{Bio} \\ B(\text{iii}) & A \in \mathcal{C}_T\mathcal{Bio}, & M \in \mathcal{C}_T A^B\text{-}\mathcal{mod} \Longrightarrow A\Pi M \in \mathcal{C}_T\mathcal{Bio} \end{array} \quad (13.6)$$

Note that in P(i), respectively, B(i), we need "more" commutativity of M, and we therefore will concentrate on $M \in \mathcal{C}_T A^P\mathcal{mod}$, respectively, $\mathcal{C}_f A^B\mathcal{mod}$.

Fixing a homomorphism $\varphi \in \mathcal{CProp}(B, A)$, respectively, $\mathcal{CBio}(B, A)$, we have the categories of objects over A, and under B, with objects

$$\mathcal{CProp}_{B\backslash\varphi/A} = \{(\varepsilon, \pi) \in \mathcal{CProp}(B, H) \times \mathcal{CProp}(H, A), \pi \circ \varepsilon = \varphi\}$$

and maps, for $H = (\varepsilon, \pi)$, $H' = (\varepsilon', \pi')$,

$$\mathcal{CProp}_{B\backslash\varphi/A}(H, H') = \{f \in \mathcal{CProp}(H, H'), f \circ \varepsilon = \varepsilon', \pi' \circ f = \pi\} \quad (13.7)$$

and similarly for Bios. We view $A \Pi M$ as an object of this category via

$$\varepsilon : B \longrightarrow A\Pi M, \quad \varepsilon(b) := \big(\varphi(b), o\big) \quad (13.8)$$

We obtain the Abelian group

$$\mathcal{P}\mathrm{Der}_B(H, M) := \mathcal{CProp}_{B\backslash\varphi/A}(H, A\Pi M)$$

$$= \left\{ \begin{array}{l} f : H \to A\Pi M, f_{n,m}(h) = (\pi_{n,m}(h), \delta_{n,m}(h)) \\ \delta_{n,m} \in \mathcal{Set}^{S_n \circ S_m^{\mathrm{op}}}(H_{n,m}, M_{n,m}) \\ \delta(h_1 \circ h_2) = \delta(h_1) \circ \pi(h_2) + \pi(h_1) \circ \delta(h_2) \\ \delta(h_1 \oplus h_2) = \delta(h_1) \oplus \delta(h_2) \\ \delta \circ \varepsilon(b) \equiv 0 \quad \text{for } b \in B \end{array} \right\} \quad (13.9)$$

and similarly,

$$\mathcal{B}\mathrm{Der}_B(H, M) := \mathcal{CBio}_{B\backslash\varphi/A}(H, A\Pi M)$$
$$= \left\{ \begin{array}{l} f : H \to A\Pi M, f_n^{\pm}(h) = \left(\pi_n^{\pm}(h), \delta_n^{\pm}(h)\right) \\ \\ \delta_n^{\pm} \in \mathcal{Set}^{S_n}(H_n^{\pm}, M_n^{\pm}) \\ \\ \delta(h_1 \circ h_2) = \delta(h_1) \circ \pi(h_2) + \pi(h_1) \circ \delta(h_2) \\ \\ \delta \circ \varepsilon(b) \equiv 0 \;\text{ for } b \in B \end{array} \right\} \tag{13.10}$$

(here $\circ$ denotes either multiplication or actions).

These give functors, taking $H = A$,

$$\begin{aligned} &\mathcal{P}\mathrm{Der}_B\left(A, \underline{?}\right) : \mathcal{C}_T A^{\mathcal{P}}\text{-}\mathcal{mod} \longrightarrow \mathcal{Ab} \\ &\mathcal{B}\mathrm{Der}_B\left(A, \underline{?}\right) : \mathcal{C}_f A^{\mathcal{B}}\text{-}\mathcal{mod} \longrightarrow \mathcal{Ab} \end{aligned} \tag{13.11}$$

These functors are representable, and we obtain

$$\begin{aligned} &\mathcal{P}\mathrm{Der}_B(A, M) \equiv \mathcal{C}_T A^{\mathcal{P}}\text{-}\mathcal{mod}(\Omega(A/B), M) \\ &\mathcal{B}\mathrm{Der}_B(A, M) \equiv \mathcal{C}_f A^{\mathcal{B}}\text{-}\mathcal{mod}(\Omega(A/B), M) \\ &f \circ d_{A/B} \longleftrightarrow f \end{aligned} \tag{13.12}$$

with the universal derivation $d_{A/B} \in \mathcal{P}\mathrm{Der}_B(A, \Omega(A/B))$, and similarly for bios.

Thus we have adjunction

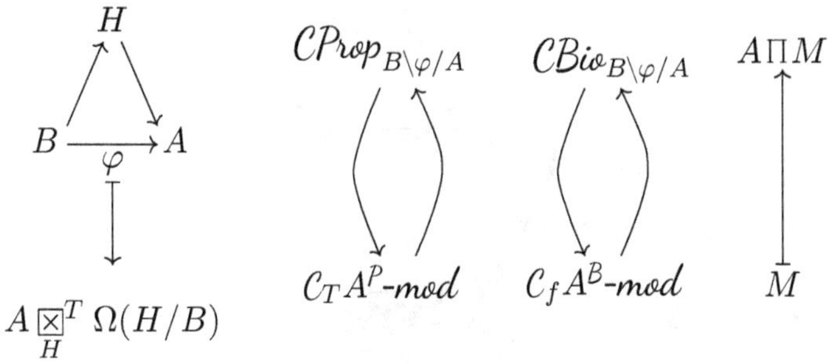

respectively, $A \underset{H}{\boxtimes}^f \Omega(H/B)$

(13.13)

For props, and similarly for bios,

$$\begin{aligned}\mathcal{CProp}_{B\backslash\varphi/A}(H, A\sqcap M) &\equiv \mathcal{P}\mathrm{Der}_B(H, \pi^* M)\\ &\equiv H^{\mathcal{P}}\text{-}\mathit{mod}(\Omega(H/B), \pi^* M)\\ &\equiv A^{\mathcal{P}}\text{-}\mathit{mod}(A \underset{H}{\boxtimes}^T \Omega(H/B), M)\end{aligned} \tag{13.14}$$

We will omit the pre-fix "$\mathcal{P}$" and "$\mathcal{B}$" treating Props and Bios simultaneously.

We will write A-mod for $\mathcal{C}_T A^{\mathcal{P}}$-$\mathit{mod}$, for a prop A, and $\mathcal{C}_f A^{\mathcal{B}}$-$\mathit{mod}$ for a bio A.

Given a commutative diagram

$$\begin{array}{ccc} A & \longrightarrow & A' \\ \uparrow & & \uparrow \\ B & \longrightarrow & B' \end{array}$$

we have homomorphisms of Abelian groups for $M' \in A'$-mod

$$\mathrm{Der}_{B'}(A', M') \longrightarrow \mathrm{Der}_B(A', M') \longrightarrow \mathrm{Der}_B(A, M') \tag{13.15}$$

represented by the A'-mod homomorphisms, and commutative diagram of derivations

$$\begin{array}{ccccc} A' \underset{A}{\boxtimes}^T \Omega(A/B) & \longrightarrow & \Omega(A'/B) & \longrightarrow & \Omega(A'/B') \\ {\scriptstyle A' \boxtimes d_{A/B}} \uparrow & & \uparrow {\scriptstyle d_{A'/B}} & \nearrow {\scriptstyle d_{A'/B'}} & \\ A & \longrightarrow & A' & & \end{array} \tag{13.16}$$

First exact sequence: Given a commutative diagram

$$\begin{array}{ccc} A & \longrightarrow & A' \\ \nwarrow & & \nearrow \\ & B & \end{array}$$

we have an exact sequence in A'-$\mathcal{mod}$

$$A' \underset{A}{\boxtimes}^T \Omega(A/B) \longrightarrow \Omega(A'/B) \longrightarrow \Omega(A'/A) \longrightarrow 0 \tag{13.17}$$

Indeed, applying A'-$\mathcal{mod}(?, M')$ this is equivalent to exactness of

$$0 \longrightarrow \mathrm{Der}_A(A', M') \longrightarrow \mathrm{Der}_B(A', M') \longrightarrow \mathrm{Der}_B(A, M')$$

which is clear from the definition.

Second exact sequence: For a surjection π

$$\ker \pi = A \underset{\widetilde{A}}{\prod} A = \mathcal{E} \rightrightarrows A \xrightarrow{\pi} \widetilde{A} = A/\mathcal{E}$$

with $B \to A$ and $B \to \widetilde{A}$ (commutative triangle).

We have an exact sequence of $\widetilde{A}$-$\mathcal{mod}$

$$\begin{gathered}\widetilde{A} \boxtimes^T \Omega(\mathcal{E}/B) \longrightarrow \widetilde{A} \boxtimes^T_A \Omega(A/B) \longrightarrow \Omega(\widetilde{A}/B) \longrightarrow 0 \\ A \underset{\widetilde{A}}{\prod} A \ni (a_1, a_2) \overset{d}{\rightsquigarrow} da_1 - da_2\end{gathered} \tag{13.18}$$

Indeed, applying $\widetilde{A}$-$\mathcal{mod}(?, \widetilde{M})$ this is equivalent to exactness of

$$\begin{gathered}0 \longrightarrow \mathrm{Der}_B(\widetilde{A}, \widetilde{M}) \longrightarrow \mathrm{Der}_B(A, \widetilde{M}) \longrightarrow \mathrm{Der}_B(\mathcal{E}, \widetilde{M}) \\ D \rightsquigarrow D(a_1, a_2) = D(a_1) - D(a_2)\end{gathered}$$

which is again clear from the definition.

Products: For $B \longrightarrow A_i$,

$$\Omega(A_1 \underset{B}{\boxtimes} A_2/B) \equiv \Big((A_1 \underset{B}{\boxtimes} A_2) \underset{A_1}{\boxtimes}^T \Omega(A_1/B)\Big) \oplus \Big((A_1 \underset{B}{\boxtimes} A_2) \underset{A_2}{\boxtimes}^T \Omega(A_2/B)\Big) \tag{13.19}$$

Base change: For push out diagram $\begin{array}{ccc} A & \rightarrow & A \underset{B}{\boxtimes} B' \\ \uparrow & & \uparrow \\ B & \longrightarrow & B' \end{array}$ we have

$$\Omega(A \underset{B}{\boxtimes} B'/B') \equiv (A \underset{B}{\boxtimes} B') \underset{A}{\boxtimes}^T \Omega(A/B) \tag{13.20}$$

Co-limits: Ω commutes with co-limits

$$\Omega(\varinjlim_{i\in I} A_i / \varinjlim_{i\in I} B_i) \equiv \varinjlim_{i\in I} \Omega(A_i/B_i) \tag{13.21}$$

Localization: For multiplicative sets:

$$\begin{aligned} &S_B \subseteq B_{1,1}\ , \quad S_A \subseteq A_{1,1}\ , \ \varphi(S_B) \subseteq S_A \\ &\Omega(S_A^{-1}A/S_B^{-1}B) \equiv S_A^{-1}\Omega(A/B) \end{aligned} \tag{13.22}$$

It follows that for a generalized-scheme map $f : X \to Y$ in $\mathcal{PSch}$, respectively, $\mathcal{BSch}$, we have

$$\Omega(X/Y) \in q.c.\mathcal{C}_T(\mathcal{O}_X)^P\text{-}\mathcal{mod}, \quad \text{respectively,} \quad q.c.\mathcal{C}_f(\mathcal{O}_X)^B\text{-}\mathcal{mod}$$

and universal derivation

$$d_{X/Y} : \mathcal{O}_X \to \Omega(X/Y) \tag{13.23}$$

Given a map $g \in \mathcal{GSch}/Y(X, X')$ of generalized-scheme over Y,we have the "First" exact sequence of $\mathcal{O}_X$-modules

$$g^*\Omega(X'/Y) \longrightarrow \Omega(X/Y) \longrightarrow \Omega(X/X') \longrightarrow 0 \tag{13.24}$$

Given a quasi-coherent $\mathcal{O}_X$-equivalence ideal $\mathcal{E} \subseteq \mathcal{O}_X \Pi \mathcal{O}_X$, writing $X = \bigcup_i \mathcal{spec} A_i$, we obtain by gluing $\widetilde{X} = \bigcup_i \mathcal{spec}(A_i/\mathcal{E}(\mathcal{spec} A_i))$, together with the immersion (not necessarily closed!)

$$j_{\mathcal{E}} : \widetilde{X} \hookrightarrow X$$

onto the "$\mathcal{E}$-stable points", and

$$j_{\mathcal{E}}^{\#} : \mathcal{O}_X \twoheadrightarrow (j_{\mathcal{E}})_* \mathcal{O}_{\widetilde{X}} \quad \text{surjective} \tag{13.25}$$

We have the "Second" exact sequence of $\mathcal{O}_{\widetilde{X}}$-modules

$$j_{\mathcal{E}}^*\Omega(\mathcal{E}/Y) \longrightarrow j_{\mathcal{E}}^*\Omega(X/Y) \longrightarrow \Omega(\widetilde{X}/Y) \longrightarrow 0 \tag{13.26}$$

Products: Denoting by $\pi_i : X_1 \prod_Y X_2 \to X_i$ the projections, we have isomorphism of $\mathcal{O}_{X_1 \prod_Y X_1}$-modules

$$\Omega\left(X_1 \prod_Y X_2 \Big/ Y\right) = \pi_1^*\Omega(X_1/Y) \oplus \pi_2^*\Omega(X_2/Y) \tag{13.27}$$

and

$$\Omega\left(X_1 \prod_Y X_2 \Big/ X_2\right) = \pi_1^*\Omega(X_1/Y) \tag{13.28}$$

Example 13.1. $\Omega(\mathbb{N}/_{\mathbb{F}})$ and $\Omega(\mathbb{Z}/_{\mathbb{F}[\pm 1]})$
These $\mathbb{N}$-modules, respectively, $\mathbb{Z}$-modules, are generated by

$$dv = d(1,1) \in \Omega_{1,2} \quad \text{and} \quad dv^t = d\begin{pmatrix}1\\1\end{pmatrix} \in \Omega_{2,1} \tag{13.29}$$

Thus, as prop-modules we will have in bi-degree (n,m) the elements

$$\left\{b\,;\, \begin{matrix}d\\d'\end{matrix}\right\} := b \circ dv \circ \begin{pmatrix}d\\d'\end{pmatrix} \quad \text{and} \quad \{b,b';d\} := (b,b') \circ dv^t \circ d$$

$$b = (b_1 \cdots b_n)^t,\ b' = (b' \cdots b'_n)^t,\ d = (d_1 \cdots d_m),\ d' = (d'_1 \cdots d'_m) \tag{13.30}$$

with $b_i, b'_i, d_j, d'_j \in \mathbb{N}$, respectively, $\mathbb{Z}$. We have the following relations:

$$\textbf{Zero:} \quad \left\{0_n^t; \begin{matrix}d\\d'\end{matrix}\right\} = \left\{b; \begin{matrix}0_m\\d'\end{matrix}\right\} = 0 = \{b,b';0_m\} = \{0_n^t, b; d\} \tag{13.31}$$

$$\textbf{Commutativity:} \quad \left\{b; \begin{matrix}d\\d'\end{matrix}\right\} = \left\{b; \begin{matrix}d'\\d\end{matrix}\right\}, \{b,b',d\} = \{b',b;d\} \tag{13.32}$$

$$\textbf{Associativity:} \quad \left\{b; \begin{matrix}d+d'\\d''\end{matrix}\right\} + \left\{b; \begin{matrix}d\\d'\end{matrix}\right\} = \left\{b; \begin{matrix}d\\d'+d''\end{matrix}\right\} + \left\{b; \begin{matrix}d'\\d''\end{matrix}\right\} \tag{13.33}$$

graphically,

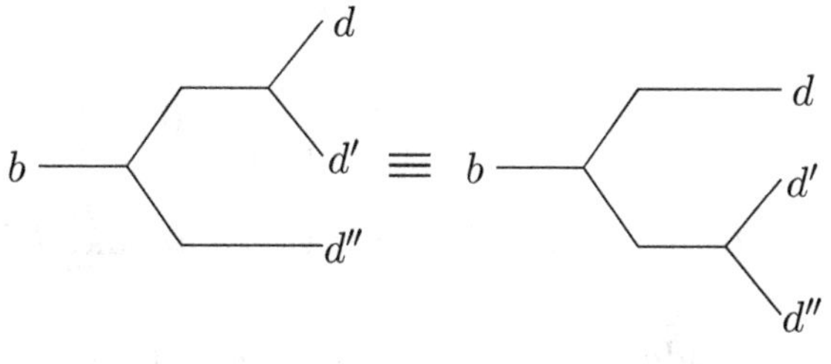

and similarly,

$$\{b+b',b'';d\}+\{b,b';d\}=\{b,b'+b'';d\}+\{b',b'';d\} \tag{13.34}$$

i.e. $\{b;\ \}$ and $\{\_,\_;d\}$ are **symmetric normalized 2-cocycles**, for fixed b,d, in their double-variables in $\mathbb{N}^m \subseteq \mathbb{Z}^m$ and $\mathbb{N}^n \subseteq \mathbb{Z}^n$ respectively.

We have further the total-commutativity of Ω, which gives in $\Omega_{2,2}$

$$(dv \oplus dv)\circ \sigma_{2,2}\circ \left(\begin{pmatrix}1\\1\end{pmatrix}\oplus\begin{pmatrix}1\\1\end{pmatrix}\right) \equiv \begin{pmatrix}1\\1\end{pmatrix}\circ dv \tag{13.35}$$

and similarly,

$$dv^t \circ (1,1) \equiv ((1,1)\oplus(1,1))\circ \sigma_{2,2}\circ (dv^t \oplus dv^t) \tag{13.36}$$

Multiplying these by $(b,b')\circ \_\_\_ \circ \begin{pmatrix}d\\d'\end{pmatrix}$ we obtain linearity in the one-variable in $\mathbb{N}^n \subseteq \mathbb{Z}^n$ and $\mathbb{N}^m \subseteq \mathbb{Z}^m$ respectively:

$$\left\{b;\begin{matrix}d\\d'\end{matrix}\right\}+\left\{b';\begin{matrix}d\\d'\end{matrix}\right\}\equiv\left\{b+b';\begin{matrix}d\\d'\end{matrix}\right\} \tag{13.37}$$

and similarly,

$$\{b,b';d+d'\}\equiv\{b,b';d\}+\{b,b';d'\} \tag{13.38}$$

For $\Omega(\mathbb{Z}/\mathbb{F}[\pm 1])$, since (-1) is central in $\mathbb{Z}$,

$$-\left\{b;\begin{matrix}d\\d'\end{matrix}\right\}=\left\{-b;\begin{matrix}d\\d'\end{matrix}\right\}=\left\{b;\begin{matrix}-d\\-d'\end{matrix}\right\} \tag{13.39}$$

and

$$-\{b,b';d\}=\{b,b';-d\}=\{-b,-b';d\} \tag{13.40}$$

We have one more relation, the "cancellation",

$$(1,1)\circ\begin{pmatrix}1&0\\0&-1\end{pmatrix}\circ\begin{pmatrix}1\\1\end{pmatrix}=0$$

which gives in $\Omega_{1,1}$ the only relation between dv and dv^t:

$$dv\circ\begin{pmatrix}1\\-1\end{pmatrix}+(1,-1)\circ dv^t=0 \tag{13.41}$$

Applying $b \circ \_ \circ d$ we obtain

$$\left\{b; \begin{matrix} d \\ -d \end{matrix}\right\} + \left\{b, -b; d\right\} = 0 \tag{13.42}$$

Note that we have

$$\left\{b; \begin{matrix} d \\ -d \end{matrix}\right\} = \left\{b; \begin{matrix} -d \\ d \end{matrix}\right\} = -\left\{b; \begin{matrix} d \\ -d \end{matrix}\right\}$$

so that the elements in (13.42) are 2-torsion.

Note that $\Omega(\mathbb{Z}/_{\mathbb{F}[\pm 1]})$ and $\Omega(\mathbb{N}/_{\mathbb{F}})$ have involution

$$\begin{aligned} &\omega \longmapsto \omega^t : \Omega_{n,m} \xrightarrow{\sim} \Omega_{m,n}, \quad \omega^{tt} = \omega \\ &(a \circ \omega \circ b)^t = b^t \circ \omega^t \circ a^t \end{aligned} \tag{13.43}$$

and this involution interchanges the generators dv and dv^t.

So let us concentrate on the abelian group

$$\Omega^+ = \Omega(\mathbb{N}/_{\mathbb{F}})^+_{1,1} = \left\{\omega \in \Omega(\mathbb{N}/_{\mathbb{F}})_{1,1}, \omega^t = \omega\right\} \tag{13.44}$$

It is generated by the symmetric elements

$$\{a, b\} = dv \circ \begin{pmatrix} a \\ b \end{pmatrix} + (a, b) \circ dv^t, \quad a, b \in \mathbb{N}^+ \tag{13.45}$$

The relations are **commutativity** (13.32)

$$\{a, b\} = \{b, a\} \tag{13.46}$$

associativity (13.33)

$$\{a + b, c\} + \{a, b\} = \{a, b + c\} + \{b, c\} \tag{13.47}$$

and from total commutativity we obtain **homogeneity**

$$\{c \cdot a, c \cdot b\} = c \cdot \{a, b\}, \quad a, b, c \in \mathbb{N}^+ \tag{13.48}$$

Thus Ω^+ is generated by $\{a, b\}$ with $a \leqslant b$, $\gcd\{a, b\} = 1$, and the one more cocycle relation (13.47), which is not consistent with this data. We have the map

$$
\begin{aligned}
& d : \mathbb{N} \longrightarrow \Omega^+ \\
\text{(i)}\ & d(0) = d(1) = 0, \quad d(n) = \{1,1\} + \{1,2\} + \cdots + \{1, n-1\} \\
\text{(ii)}\ & d(n \cdot m) = d(n) \cdot m + n \cdot d(m) \\
\text{(iii)}\ & d(n+m) = d(n) + d(m) + \{n, m\} \\
\text{(iv)}\ & d(n) = \textstyle\sum_p v_p(n) \cdot \frac{n}{p} d(p) \\
& \text{or the “dlog” formula in } \Omega^+_{\mathbb{Q}} := \Omega^+ \otimes \mathbb{Q} \equiv \Omega(\mathbb{Q}^+/_{\mathbb{F}})^+_{1,1} \\
\text{(v)}\ & \tfrac{d(n)}{n} = \textstyle\sum_p v_p(n) \cdot \frac{dp}{p}, \ n \in \mathbb{Q}^+.
\end{aligned}
\tag{13.49}
$$

Note that for a prime p we have the well defined homomorphisms

$$
\varphi_p : \Omega^+ \longrightarrow \mathbb{Z}, \ \varphi_p(\{a,b\}) = v_p(a+b) \cdot \frac{(a+b)}{p} - v_p(a) \cdot \frac{a}{p} - v_p(b) \cdot \frac{b}{p} \in \mathbb{Z}
\tag{13.50}
$$

Indeed one check that it satisfies the relations (13.46), (13.47), (13.48). It follows that Ω^+ is the free abelian group on the primes

$$
\Omega^+ \cong \bigoplus_p \mathbb{Z} d(p), \quad d(p) = \{1,1\} + \cdots + \{1, p-1\}
\tag{13.51}
$$

We can extend d to $d : \mathbb{Q} \to \Omega^+_{\mathbb{Q}}$, by $d(-n) = -d(n)$, and rewrite (13.49) (v) as

$$
|n|_{\mathbb{R}} \equiv e^{\int \frac{d(n)}{n}}, \quad n \in \mathbb{Q}^*
\tag{13.52}
$$

$\int : \Omega^+_{\mathbb{Q}} \hookrightarrow \mathbb{R}$ the linear map given by $\int \frac{d(p)}{p} = \log p$, for p prime.

Similarly extending scalars to $\mathbb{Z}$ we get the “vector fields”

$$
\partial_p : \mathbb{Z} \to \mathbb{Z}, \quad \partial_p(\pm n) = \pm \partial_p(n) = \pm \varphi_p \circ d(n) = v_p(n) \cdot \frac{(\pm n)}{p}
\tag{13.53}
$$

and the “global differentiation”

$$
\partial = \sum_p \partial_p : \mathbb{Z} \to \mathbb{Z}, \quad \partial(n) = \sum_p v_p(n) \cdot \frac{n}{p}, \quad n \in \mathbb{Z}
\tag{13.54}
$$

Chapter 14

Simplicial Objects and the Cotangent Complex

We denote by $\triangle$ the simplicial category of finite ordered sets, with objects

$$[n] = \{0 \leqslant 1 \leqslant \cdots \leqslant n\}, \quad n \geqslant 0 \tag{14.1}$$

and with order preserving functions. For a category $\mathcal{C}$, we let $\mathcal{S}\mathcal{C} = (\mathcal{C})^{\triangle^{\mathrm{op}}}$, the category of contravariant functors $\triangle^{\mathrm{op}} \longrightarrow \mathcal{C}$, $[n] \rightsquigarrow A^n$, $\triangle(n,m) \longrightarrow \mathcal{C}(A^m, A^n)$. We use **upper indices** for the simplicial dimensions (since for props the lower indices are already used).

The categories $\mathcal{S}\mathcal{C}_f\mathcal{B}io$ and $\mathcal{S}\mathcal{C}_f\mathcal{P}rop$ have the **projective Quillen model structures** with

Fibration:

$$\mathcal{F}ib = \{\varphi^* : B^* \to A^*, \forall n, m \{\varphi^*_{n,m} : B^*_{n,m} \to A^*_{n,m}\} \in \mathcal{F}ib_{\mathcal{S}et_0}\}$$

equivalence:

$$\mathcal{W} = \{\varphi^* : B^* \to A^*, \forall n, m \{\varphi^*_{n,m} : B^*_{n,m} \to A^*_{n,m}\} \in \mathcal{W}_{\mathcal{S}et_0}\} \tag{14.2}$$

the fibrations/weak equivalence of pointed simplicial sets for all n, m (for props; for bios: for all n and $\epsilon \in \{\pm\}$). The cofibration, the maps having the left-lifting-property with respect to all acyclic-fibrations

$$\mathcal{C}of = \mathcal{L}(\mathcal{W} \cap \mathcal{F}ib) \tag{14.3}$$

are the retracts of free-maps. A map $\varphi^* : B^* \to A^*$ is **free** if there exists subsets $V^n \subseteq A^n$ (not necessarily of (bi)degree 1!) such that

$$\ell^*(V^m) \subseteq V^n \text{ for all } \ell \in \triangle_{\text{surj}}([n],[m]) \text{ surjective} \tag{14.4}$$

and φ^n induces isomorphism

$$B^n \boxtimes^f \mathbb{F}[V^n] \xrightarrow{\sim} A^n \tag{14.5}$$

with $\mathbb{F}[V^n]$ the free $\mathcal{C}_f\mathcal{B}io$ or $\mathcal{C}_f\mathcal{P}rop$ on V^n (and the tensor product taken in $\mathcal{C}_f\mathcal{B}io$ or $\mathcal{C}_f\mathcal{P}rop$).

This model structure is cofibrantly generated, and the (trivial) fibrations are the maps satisfying the right-lifting-property with respect to the canonical maps

$$\begin{aligned} \mathcal{W} \cap \mathcal{F}ib &\equiv \mathcal{R}\{\mathbb{F}[\partial\Delta(d)]_{n,m} \to \mathbb{F}[\Delta(d)]_{n,m},\ d > 0,\ n,m \geqslant 0\} \\ \mathcal{F}ib &\equiv \mathcal{R}\{\mathbb{F}[\Lambda^d_k]_{n,m} \to \mathbb{F}[\Delta(d)]_{n,m},\ 0 \leqslant k \leqslant d > 0,\ n,m \geqslant 0\} \end{aligned} \tag{14.6}$$

where for a (simplicial) set V, and $n,m \geqslant 1$ (or $n \geqslant 1$, $\epsilon = \pm$) we let $\mathbb{F}[V]_{n,m}$ (or $\mathbb{F}[V^\epsilon(n)]$), denote the free (simplicial) $\mathcal{C}_f\mathcal{P}rop$, (or $\mathcal{C}_f\mathcal{B}io$), generated by V in bi-degree n,m (n,ϵ)

$$\mathcal{S}\mathcal{C}_f\mathcal{P}rop(\mathbb{F}[V]_{n,m}, A^*) \equiv \mathcal{S}\mathcal{S}et(V, A^*_{n,m}) \tag{14.7}$$

(and similarly for $\mathcal{S}\mathcal{C}_f\mathcal{B}io$).

For $A^* \in \mathcal{S}\mathcal{C}_f\mathcal{P}rop$ (respectively, $\mathcal{S}\mathcal{C}_f\mathcal{B}io$) and a nice category $\mathcal{C}$, we have the categories

$$\mathcal{S}A^{*P}[\mathcal{C}], \quad \text{respectively,} \quad \mathcal{S}A^{*B}[\mathcal{C}] \quad \text{or} \quad \mathcal{S}A^{*1}[\mathcal{C}] \tag{14.8}$$

of simplicial prop (respectively, bio or one) objects of $\mathcal{C}$, consisting of $M^* = \{M^d\}_{d\geqslant 0}$, with M^d a prop (respectively, bio or one) A^d-object of $\mathcal{C}$, and for $\ell \in \triangle([n],[m])$

$$\ell^* : M^m \longrightarrow M^n \tag{14.9}$$

a map compatible with $\ell^* : A^m \to A^n$,

$$(\ell_2 \circ \ell_1)^* = \ell_1^* \circ \ell_2^*, \quad (\mathrm{Id}_{[n]})^* = \mathrm{Id}_{M^n} \tag{14.10}$$

We will be mainly interested in the categories $\mathcal{S}A^{*\mathcal{P}}[\mathcal{Ab}]$, or $\mathcal{S}A^{*\mathcal{B}}[\mathcal{Ab}]$, we will write $\mathcal{S}A^*\text{-}\mathcal{mod}$ for these; and the category

$$\mathcal{S}A^*\text{-}\mathcal{Set} \equiv \mathcal{S}A^{*1}[\mathcal{Set}]$$

These categories have the **projective Quillen model structure** with

Fibrations:

$$\mathcal{Fib} \equiv \{\varphi^* : M^* \to N^*, \{\varphi^*_{n,m} : M^*_{n,m} \to N^*_{n,m}\} \in \mathcal{Fib}_{\mathcal{Set}_0},\ \text{all } n, m\}$$

Weak equivalence:

$$\mathcal{W} \equiv \{\varphi^* : M^* \to N^*, \{\varphi^*_{n,m} : M^*_{n,m} \to N^*_{n,m}\} \in \mathcal{W}_{\mathcal{Set}_0},\ \text{all } n, m\}$$

Cofibratioons: $\mathcal{Cof} \equiv \mathcal{L}(\mathcal{W} \cap \mathcal{Fib})$ (14.11)

for $\mathcal{Props}$ and similarly for $\mathcal{Bios}$.

Again the cofibrations are the retracts of **free-maps**, where $\varphi^* : M^* \to N^*$ is **free** if there are subsets $V^n \subseteq N^n$ such that

$$\ell^*(V^m) \subseteq V^n \ \text{ for all } \ell \in \triangle_{\text{surj}}([n],[m]) \text{ surjective} \tag{14.12}$$

and φ^n induces isomorphism

$$M^n \amalg A^n \bullet V^n \xrightarrow{\sim} N^n \tag{14.13}$$

with $A^n \bullet V^n$ the free A^n-module, or A^n-$\mathcal{Set}$, on V^n.

Again this model structure is cofibrantly-generated

$$\mathcal{W} \cap \mathcal{Fib} \equiv \mathcal{R}\{A^* \bullet \partial\Delta(d)_{n,m} \to A^* \bullet \Delta(d)_{n,m},\ \ d, n, m > 0\}$$
$$\mathcal{Fib} \equiv \mathcal{R}\{A^* \bullet (\Lambda^d_k)_{n,m} \to A^* \bullet \Delta(d)_{n,m},\ \ 0 \leqslant k \leqslant d, n, m > 0\} \tag{14.14}$$

For A^*-$\mathcal{mod}$ we have under the Dold-Kan equivalence

$$\mathcal{S}A^*\text{-}\mathcal{mod} \xleftrightarrow{\sim} Ch_{\geqslant 0}(A^*\text{-}\mathcal{mod}) \tag{14.15}$$

and we can realize this model structure using chain complexes

$$M^* \equiv \{0 \longleftarrow M^0 \xleftarrow{\partial} M^1 \xleftarrow{\partial} \cdots \xleftarrow{\partial} M^d \xleftarrow{\partial} M^{d+1} \xleftarrow{\partial} \cdots\},$$
$$\partial \circ \partial = 0 \tag{14.16}$$

with M^d an A^d-module, using the projective model structure on chain complexes:

Fibrations:

$$\mathcal{Fib} = \{\varphi^* : M^* \to N^*, \varphi^d : M^d \twoheadrightarrow N^d \text{ surjective } d \geqslant 0\}$$

Weak equivalence $\equiv$ quasi-isomorphisms:

$$\mathcal{W} \equiv \{\varphi^* : M^* \to N^*, H_d(\varphi) : H_d(M^*) \xrightarrow{\sim} H_d(N^*) \text{ is an isomorphism } d \geqslant 0\}$$

Cofibrations:

$$\mathcal{Cof} \equiv \{\varphi^* : M^* \to N^*, \varphi^d \text{ injective with Coker } \varphi^d \text{ projective}\} \tag{14.17}$$

For a free $\mathcal{C}_f\mathcal{Prop}$ or $\mathcal{C}_f\mathcal{Bio}$ under B, the "polynomials" on variables of multi-degrees,

$$A = B \boxtimes^f \mathbb{F}[x_i] \equiv B[x_i], \quad x_i \text{ (bi)-degree } (n_i, m_i) \text{ or } (n_i, \pm) \tag{14.18}$$

we have

$$\Omega(A/B) \equiv A \bullet \{dx_i\} \quad \text{the free } A\text{-module on } \{dx_i\} \tag{14.19}$$

Thus the left adjoint functor of (12.13)

$$\begin{array}{ccc} & H & \\ \nearrow & & \searrow \\ B & \longrightarrow & A \end{array} \quad \rightsquigarrow \quad A \underset{H}{\boxtimes}^f \Omega(H/B)$$

takes (acyclic) cofibrations to (acyclic) cofibrations, and is a left Quillen functor.

For a homomorphism of (simplicial) $\mathcal{S}\mathcal{C}_f\mathcal{Prop}$ or $\mathcal{S}\mathcal{C}_f\mathcal{Bio}$, $\varphi^* : B^* \to A^*$ the **cotangent-complex** $\mathbb{L}\Omega(A^*/B^*)$ is the element of the derived category of A-$\mathcal{mod}$, A-$\mathcal{mod}[\mathcal{W}^{-1}] := \mathbb{D}(A\text{-}\mathcal{mod})$, given by

$$\mathbb{L}\Omega(A^*/B^*) := \Omega\big(P_{B*}(A^*)/B^*\big) \underset{P_{B*}(A^*)}{\boxtimes}^f A^* \in \mathbb{D}(A^*\text{-}\mathcal{mod})$$

$$\text{where } P_{B*}(A^*) \longrightarrow A^* \quad \text{is a cofibrant replacement of } A^* \tag{14.20}$$

The derived category of A-$\mathcal{mod}$ is given explicitly as

$$\begin{aligned}\mathbb{D}(A^*\text{-}\mathcal{mod}) &\equiv \mathrm{Ho}(\mathcal{S}A^*\text{-}\mathcal{mod})\\ &\cong \mathrm{Ch}_{d\geqslant 0}(\text{projective-}A^d\text{-}\mathcal{mod})/\text{homotopy}\end{aligned}$$

Given a commutative diagram of homomorphisms

$$\begin{array}{ccccc} B^* & \longrightarrow & A^* & \longrightarrow & \overline{A}^* \\ \| & & \uparrow & & \uparrow \\ B^* & \longrightarrow & Q^* \equiv P_{B^*}(A^*) & \longrightarrow & P_{Q^*}(\overline{A}^*) \end{array}$$

we have compatible cofibrant replacements, and we have a short exact sequence of $\overline{A}^*$-modules (also on the left!)

$$0 \longrightarrow \overline{A}^* \underset{Q^*}{\boxtimes} \Omega(Q^*/B^*) \longrightarrow \overline{A}^* \underset{P^*_{Q^*}(\overline{A}^*)}{\boxtimes} \Omega\left(P_{Q^*}(\overline{A}^*)/B^*\right)$$
$$\longrightarrow \overline{A}^* \underset{P^*_{Q^*}(\overline{A}^*)}{\boxtimes} \Omega\left(P_{Q^*}(\overline{A}^*)/Q^*\right) \longrightarrow 0 \tag{14.21}$$

Taking homology we get the long exact sequence of $\overline{A}^*$-modules (extending the first exact sequence (12.17))

$$\overline{A}^* \boxtimes_{A^*} \Omega(A^*/B^*) \longrightarrow \Omega(\overline{A}^*/B^*) \longrightarrow \Omega(\overline{A}^*/A^*) \longrightarrow 0$$
$$\overline{A}^* \boxtimes_{A^*} L_1\Omega(A^*/B^*) \longrightarrow L_1\Omega(\overline{A}^*/B^*) \longrightarrow L_1\Omega(\overline{A}^*/A^*)$$
$$\cdots \overline{A}^* \boxtimes_{A^*} L_2\Omega(A^*/B^*) \longrightarrow L_2\Omega(\overline{A}^*/B^*) \longrightarrow L_2\Omega(\overline{A}^*/A^*)$$

$$L_n\Omega(A^*/B^*) \equiv H_n(A^* \underset{Q^*}{\boxtimes} \Omega(Q^*/B^*)) \equiv \frac{\mathrm{Ker}\{A\boxtimes\Omega(Q^n/B)\to A\boxtimes\Omega(Q^{n-1}/B)\}}{\mathrm{Image}\{A\boxtimes\Omega(Q^{n+1}/B)\to A\boxtimes\Omega(Q^n/B)\}}$$

$$Q^* = P_{B^*}(A^*) \to A^* \quad \text{a cofibrant replacement} \tag{14.22}$$

This can be globalize: for a map of generalized schemes $f : X \to Y$, the cotangent complex $\mathbb{L}\Omega(X/Y)$ is the element of the derived category of quasi-coherent $\mathcal{O}_X$-modules,

$$\mathbb{L}\Omega(X/Y) := \mathcal{O}_X \underset{Q^*}{\boxtimes} \Omega(Q^*/f^*\mathcal{O}_Y) \in \mathbb{D}(\text{q.c. } \mathcal{O}_X\text{-}\mathit{mod}) \tag{14.23}$$

$$Q^* \longrightarrow \mathcal{O}_X \quad \text{an} \quad f^*\mathcal{O}_Y\text{-cofibrant replacement}$$

Given another map of generalized schemes $g : Y \to Z$, we get the exact triangle

$$\mathcal{O}_X \boxtimes \mathbb{L}\Omega(Y/Z) \longrightarrow \mathbb{L}\Omega(X/Z) \longrightarrow \mathbb{L}\Omega(X/Y) \tag{14.24}$$

Taking homology we get a long exact sequence of $\mathcal{O}_X$-module (extending the sequence appearing in (12.24)):

$$f^*\Omega(Y/Z) \longrightarrow \Omega(X/Z) \longrightarrow \Omega(X/Y) \longrightarrow 0$$

$$f^*L_1\Omega(Y/Z) \longrightarrow L_1\Omega(X/Z) \longrightarrow L_1\Omega(X/Y)$$

$$\cdots \longrightarrow f^*L_2\Omega(Y/Z) \longrightarrow L_2\Omega(X/Z) \longrightarrow L_2\Omega(X/Y)$$

Chapter 15

Properties of Generalized Schemes

Definition 15.1. A generalized scheme $(X, \mathcal{O}_X)$ is **irreducible** if X is topological irreducible (for $\emptyset \neq \mathcal{U}_i \subseteq X$ open, $\mathcal{U}_1 \cap \mathcal{U}_2 \neq \emptyset$). Equivalently, for every open affine $\mathcal{S}pec A \subseteq X$, $\mathcal{S}pec A$ is irreducible: $\sqrt{0}$ is a prime ideal.

Definition 15.2. A generalized scheme $(X, \mathcal{O}_X)$ is **reduced** if for all open $\mathcal{U} \subseteq X$, $\mathcal{O}_X(\mathcal{U})$ has no nilpotent.
Equivalently, for every open affine $\mathcal{S}pec A \subseteq X$, A has no nilpotent: $\sqrt{0} = 0$.

Definition 15.3. A generalized scheme $(X, \mathcal{O}_X)$ is **integral** if

for all open $\mathcal{U} \subseteq X$, $\mathcal{O}_X(\mathcal{U})$ is a domain:

no zero divisors $\Leftrightarrow$ (0) is a prime.

Equivalently, X is integral iff X is reduced and irreducible.

If g_X is the generic point of X, $K = \mathcal{O}_{X,g_x}$ is a field, the **field of rational functions** on X. Equivalently, for any open affine

$$\mathcal{S}pec A \subseteq X,$$

$K = (A(1) \backslash \{0\})^{-1} A$ is the **fraction field** of the domain A.

In this case, we have embeddings of the stalks $\mathcal{O}_{X,x} \subseteq K$, $x \in X$, and for open $\mathcal{U} \subseteq X$,

$$\mathcal{O}_X(\mathcal{U}) \equiv \bigcap_{x \in \mathcal{U}} \mathcal{O}_{X,x} \subseteq K \tag{15.1}$$

We have the adjunctions between commutative props and bios and commutative monoids $\mathcal{CM}on$

$$\begin{array}{ccccc} \mathbb{F}[M] & \mathcal{CP}rop & & \mathcal{CB}io & A \\ \uparrow & \uparrow\downarrow & & \uparrow\downarrow & \downarrow \\ M & \mathcal{CM}on & & \mathcal{CM}on & A_{1,1}, A(1) \end{array} \tag{15.2}$$

Here as a prop

$$\mathbb{F}[M]_{n,m} \equiv \left\{ \begin{array}{l} \text{n by m matrices with values in } M \amalg \{0\} \\ \text{such that every row/column has at most} \\ \text{one} \neq 0 \text{ term} \end{array} \right\} \tag{15.3}$$

As a bio, $\mathbb{F}[M] := U\mathbb{F}[M]$, and

$$\mathbb{F}[M]^{\pm}(n) \equiv \{0\} \amalg (M \times \{1, \ldots, n\}) \tag{15.4}$$

For M commutative, the prop/bio $\mathbb{F}[M]$ is totally commutative.

Taking for M the free commutative monoid on k generators $M = X_1^{\mathbb{N}} \cdots X_k^{\mathbb{N}}$ we get the "polynomial prop/bio":

$$\mathbb{F}[X_1, \ldots, X_k] := \mathbb{F}[X_1^{\mathbb{N}} \cdots X_k^{\mathbb{N}}] \tag{15.5}$$

$$\mathcal{CP}rop(\mathbb{F}[X_1 \cdots X_k], A) \equiv A_{1,1}^k, \quad \mathcal{CB}io(\mathbb{F}[X_1 \cdots X_k], A) \equiv A(1)^k$$

We will write $\varphi : B \to A$ for either

$$\varphi \in \mathcal{C}_f\mathcal{P}rop(B, A) \text{ or } \varphi \in \mathcal{C}_f\mathcal{B}io(B, A).$$

Definition 15.4. We say $\varphi : B \to A$ is of **finite type** if there is a finite set $a_1, \ldots, a_n$ of (bi-)degree 1, $a_i \in A(1)$ or $A_{1,1}$, which generate A over B,

$$A = B[a_1, \ldots, a_N] \tag{15.6}$$

Equivalently, we have a surjection

$$\begin{aligned} B[X_1, \cdot, X_N] := B \boxtimes^f \mathbb{F}[X_1, \ldots, X_N] &\twoheadrightarrow A \\ X_i &\rightsquigarrow a_i \end{aligned} \tag{15.7}$$

Note that we allow generators only in (bi-) degree 1, so that $\mathbb{N}$, and $\mathbb{Z}$, are **not** of finite type over $\mathbb{F}$, but they are finitely presented in the following sense.

A is **finitely generated** as a B prop or bio if there are finitely many generator (of arbitrary degrees), or equivalently,

$$\varinjlim_n \mathcal{C}_f\mathcal{Prop}_{B\backslash}(A, C_n) \twoheadrightarrow \mathcal{C}_f\mathcal{Prop}_{B\backslash}\left(A, \varinjlim_n C_n\right) \tag{15.8}$$

is surjective, and similarly for bios. A is **finitely presented** as B prop or bio if the relations between these generators are finitely generated, or equivalently, the map in (15.8) is a bijection.

Thus $\mathbb{Z}/\mathbb{F}$ and $\mathbb{N}/\mathbb{F}$ are finitely presented, but not of finite type.

Remark that for $B \in \mathcal{C}_f\mathcal{Bio}$, every element of $B[X_1 \cdots X_N]^-(m)$ (respectively, $B[X_1 \cdots X_N]^+(m)$), when the X_i have degree one, can be represented as

$$b \circ (X^{e_k}) \circ (d_1, \ldots, d_m), \quad \text{(respectively, } (b_1, \ldots, b_m) \circ (X^{e_k}) \circ d) \tag{15.9}$$

with $b \in B^-(\sum_{i=1}^m n_i)$, $d_i \in B^+(n_i)$,
(respectively, $b_i \in B^-(n_i)$, $d \in B^+(\sum_{i=1}^m n_i)$) and

$$X^{e_k} := X_1^{e_{k,1}} \circ \cdots \circ X_N^{e_{k,N}} \quad \text{for } k = 1, \ldots, \sum_{i=1}^m n_i$$

We define the following classes of maps of generalized schemes

$$f \in \mathcal{GSch}, \quad \text{with } G = \mathcal{P}_f \text{ or } \mathcal{B}_f \tag{15.10}$$

(i) **(open)** $\ni f$ is an **open immersion** if it is (isomorphic to) an embedding of an open subset $\mathcal{U} \subseteq Y$, $f : X \xrightarrow{\sim} \mathcal{U} \subseteq Y$, $f_*\mathcal{O}_X = \mathcal{O}_Y|_{\mathcal{U}}$.

(ii) **(loc.f.typ)** $\ni f$ is **locally of finite type** if $Y = \bigcup_i \mathit{specB}_i$ open affine cover, with $f^{-1}(\mathit{specB}_i) = \bigcup_j \mathit{specA}_{i,j}$, and $A_{i,j}/B_i$ of finite type.

(iii) **(q.comp)** $\ni f$ is **quasi-compact** iff^{-1}(compact) = compact, or equivalently, $Y = \bigcup_i \mathit{specB}_i$, and

$$f^{-1}(\mathit{specB}_i) = \bigcup_{j=1}^{N_i} \mathit{specA}_{i,j}$$

with N_i finite.

(iv) **(f.typ)** $\equiv$ (loc.f.typ) $\cap$ (*q.comp*) the maps of **finite type**.

(v) **(aff)** $\ni f$ is **affine** if $Y = \bigcup_i \mathit{specB}_i$, and $f^{-1}(\mathit{specB}_i) \equiv \mathit{specA}_i$ is affine.

(vi) **(q.sep)** $\ni f$ is **quasi-separated** if

$$Y = \bigcup_i \mathit{specB}_i, \ f^{-1}(\mathit{specB}_i) = \bigcup_j \mathit{specA}_{i,j}$$

and

$$\mathit{specA}_{i,j_1} \cap \mathit{specA}_{i,j_2} = \bigcup_{k=1}^{N_{i,j_1,j_2}} \mathit{specA}_{i,j_1,j_2,k}$$

are compact: $N_{i,j_1,j_2} < \infty$.

(vii) **(q-coh)** $\equiv$ (q.comp) $\cap$ (q.sep) $\ni f$ is **quasi-coherent** if

$$Y = \bigcup_i \mathit{specB}_i, \quad f^{-1}(\mathit{specB}_i) = \bigcup_{j=1}^{N_i} \mathit{specA}_{i,j}$$

and

$$\mathit{specA}_{i,j_1} \cap \mathit{specA}_{i,j_2} = \bigcup_{k=1}^{N_{i,j_1,j_2}} \mathit{specA}_{i,j_1,j_2,k}$$

(viii) **(u.r.)** $\ni f$ is **unramified** if it is of finite type and $\Omega(X/Y) \equiv 0$.

For "flatness" we have a variety of choices, we only hint on different possibilities.

(ix) (**flat**) $\ni f$ is **flat** if $X = \bigcup_i \mathit{spec} B_i$, $f^{-1}(\mathit{spec} B_i) = \bigcup_j \mathit{spec} A_{i,j}$, and $A_{i,j}/B_i$ "flat", where for a map $\varphi : B \to A$ we have the various adjunctions

$$\begin{array}{ccccc} A\text{-}\mathit{mod} \equiv \mathcal{C}_T A^P[\mathcal{Ab}], & & & A\text{-}\mathcal{Set} \equiv \mathcal{C}_T A^1[\mathcal{Set}], & \\ \text{respectively, } \mathcal{C}_f A^B[\mathcal{Ab}] & & & \text{respectively, } \mathcal{C}_f A^1[\mathcal{Set}] & \\ \varphi_*^{\mathcal{Ab}} \uparrow \quad \downarrow \varphi^* & & \cdots & \varphi_*^{\mathcal{Set}} \uparrow \quad \downarrow \varphi^* & \\ B\text{-}\mathit{mod} \equiv \mathcal{C}_T B^P[\mathcal{Ab}], & & & B\text{-}\mathcal{Set} \equiv \mathcal{C}_T B^1[\mathcal{Set}], & \\ \text{respectively, } \mathcal{C}_f B^B[\mathcal{Ab}] & & & \text{respectively, } \mathcal{C}_f B^1[\mathcal{Set}] & \end{array} \tag{15.11}$$

and A/B "flat" can mean a variety of choices, such as $\varphi_*^{\mathcal{Ab}}$ left exact and/or $\varphi_*^{\mathcal{Set}}$ preserves finite limits.
Unfortunately, flatness is not preserved automatically by base changed, and so when we say A/B is flat in any sense we mean that for all homomorphism $\varphi : B \to B'$, also $A \underset{B}{\boxtimes}^f B'/B'$ is flat in that sense.
For Example, Any localization $A = S^{-1}B/B$ is flat in all senses.

(x) (**ét**) = (flat) $\cap$ (u.r.) $\ni f$ is **étale** if it is flat in all senses and unramified.

Remark 15.1. Above properties are local on Y, and are equivalent to asking the same thing for any open affine $\mathit{spec} B \subseteq Y$.

All the above properties are preserved under base change (for flatness by definition) and under compositions.

We have the following implications:

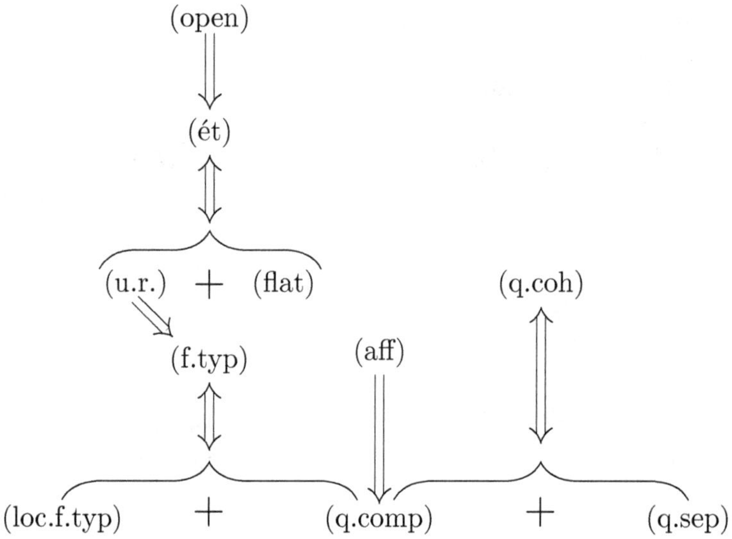
(open)
(ét)
(u.r.) + (flat)
(q.coh)
(f.typ)
(aff)
(loc.f.typ) + (q.comp) + (q.sep)

(15.12)

Remark 15.2. For a quasi-coherent map of generalized schemes $f : X \to Y$, the functor f_* (left adjoint to f^*) takes quasi-coherent sheaves M on X to quasi coherent sheaves f_*M on Y. Indeed, using the notations of ((15.10) vii), we have that $f_*M\big|_{\mathit{spec}B_i}$ is the equalizer

$$f_*M\big|_{\mathit{spec}B_i} \longrightarrow \prod_{j=1}^{N_i} f_*(M\big|_{\mathit{spec}A_{i,j}}) \Longrightarrow \prod_{\substack{j_1,j_2\leqslant N_i\\ k\leqslant N_{i,j_1,j_2}}} f_*(M\big|_{\mathit{spec}A_{i,j_1,j_2,k}}) \tag{15.13}$$

and

$$f_*(M\big|_{\mathit{spec}A_{i,j}}) = \left(f_i^{\#}M(\mathit{spec}A_i)\right)^{\sim},$$

$$f_*(M\big|_{\mathit{spec}A_{i,j_1,j_2,k}}) = \left(f^{\#}M(\mathit{spec}A_{i,j_1,j_2,k})\right)^{\sim}$$

are quasi-coherent, and so f_*M is quasi-coherent. Thus for a quasi-coherent map $f : X \to Y$ we have the adjunctions for nice categories $\mathcal{C}$

$$\mathrm{q.c.}\ \mathcal{O}_Y[\mathcal{C}] \ \underset{f_*}{\overset{f^*}{\rightleftarrows}}\ \mathrm{q.c.}\ \mathcal{O}_X[\mathcal{C}] \tag{15.14}$$

Remark 15.3. We get the various **flat** (respectively, **étale**) topologies on props and bios, taking our maps to be finitely-presented, quasi-coherent, flat (respectively, étale) maps, and the coverings $\{B \to A_i\}$ are finite collections of maps that are conservative or faithful on $\mathcal{C}_f\mathcal{P}rop_{B\backslash}$ or $\mathcal{C}_f\mathcal{B}io_{B\backslash}$ and/or on B-$\mathcal{mod}$, and/or on B-$\mathcal{S}et$, so that $C \to D$ is an isomorphism in $\mathcal{C}_f\mathcal{P}rop_{B\backslash}\big/ B\text{-}\mathcal{mod}\big/ B\text{-}\mathcal{S}et$ iff $A_i \underset{B}{\boxtimes} C \to A_i \underset{B}{\boxtimes} D$ is an isomorphism for all i.

Chapter 16

Higher K-Theory

For $A \in \mathcal{Prop}$ not necessarily commutative!, we have the groups $\mathrm{GL}_n(A)$, forming the symmetric monoidal category

$$\begin{aligned} &\mathrm{GL}(A) := \coprod_{n \geqslant 0} \mathrm{GL}_n(A) \\ &\oplus : \mathrm{GL}(A) \times \mathrm{GL}(A) \longrightarrow \mathrm{GL}(A) \\ &\text{on objects: } n_1 \oplus n_2 \ := \ n_1 + n_2 \\ &\text{on arrows: } \quad g_1 \oplus g_2 \end{aligned} \tag{16.1}$$

Here $\mathrm{GL}_0(A) = \{\mathrm{id}_0\}$, is the (strict) unit, and note that translations are faithful ($g_i \equiv I_i \circ (g_1 \oplus g_2) \circ I_i^t$).

We can define the higher K-groups of A either using the $+$-construction $\mathrm{BGL}_\infty(A)^+$, or as "the group completion" of $\mathrm{BGL}(A)$, or using Segal's construction [**Seg74**] or more naturally and canonically using a variant of Quillen's $S^{-1}S$ construction which we describe next.

We construct a symmetric monoidal category $\mathcal{K}_A$ following Quillen with objects $\mathbb{N} \times \mathbb{N}$ and maps

$$\mathcal{K}_A\left((m_1, m_2), (n_1, n_2)\right) \equiv \begin{cases} \mathrm{GL}_{n_1}(A) \times \mathrm{GL}_{n_2}(A) \Big/ \{(I_{m_1} \oplus g, I_{m_2} \oplus g), g \in \mathrm{GL}_{n_i - m_i}(A)\} \\ \qquad\qquad\qquad\qquad \text{if } n_1 - m_1 = n_2 - m_2 \geqslant 0 \\ \varnothing \qquad\qquad\qquad\qquad \text{otherwise} \end{cases} \tag{16.2}$$

For $(g_1, g_2)/_\sim \in \mathcal{K}_A((m_1, m_2), (n_1, n_2))$ and $(f_1, f_2)/_\sim \in \mathcal{K}_A((n_1, n_2), (k_1, k_2))$ we have their well-defined composition

$$(f_1, f_2)/_\sim \circ (g_1, g_2)/_\sim :\equiv \left(f_1 \circ (g_1 \oplus I_{k_1 - n_1}), f_2 \circ (g_2 \oplus I_{k_2 - n_2})\right) \Big/_\sim \tag{16.3}$$

making $\mathcal{K}_A$ into a category. Moreover, $\mathcal{K}_A$ is symmetric monoidal via

$$\begin{aligned} \bigoplus : \quad & \mathcal{K}_A \times \mathcal{K}_A \longrightarrow \mathcal{K}_A \\ \text{on objects}: \quad & (m_1, m_2) \oplus (n_1, n_2) = (m_1 + n_1, m_2 + n_2) \\ \text{on arrows}: \quad & (f_1, f_2)/_\sim \oplus (g_1, g_2)/_\sim = (f_1 \oplus g_1, f_2 \oplus g_2)/_\sim \end{aligned} \tag{16.4}$$

We have a (strict) symmetric monoidal functor

$$\begin{aligned} : \mathrm{GL}(A) & \hookrightarrow \mathcal{K}_A \\ \text{on objects}: n & \longmapsto (n, 0) \\ \text{on arrows}: g & \longmapsto (g, \mathrm{id}_0)/_\sim \end{aligned} \tag{16.5}$$

The higher K-groups of A are the homotopy groups of the infinite loop space $B\mathcal{K}_A$, the geometric realization of $\mathcal{K}_A$,

$$K_n(A) := \pi_n(B\mathcal{K}_A) \tag{16.6}$$

The classifying space functor B gives the map

$$\mathrm{BGL}(A) \longrightarrow B\mathcal{K}_A \equiv \mathbb{K}(A) \tag{16.7}$$

and this is a group completion of the H-space $\mathrm{BGL}(A)$. We will think of $B\mathcal{K}_A$ as a (symmetric) spectra and denote it by $\mathbb{K}(A)$.

Examples 16.1. For a commutative ring A the groups $K_n(A)$ agree with Quillen's higher K-groups [**Gra76**].

For the initial object $\mathbb{F}$ we have by [**Qui69**], [**BP72**],

$$K_n(\mathbb{F}) \equiv \pi_n^{\mathrm{st}} \equiv \lim_{m \to \infty} \pi_{n+m}(S^m) \tag{16.8}$$

are the stable homotopy groups of the spheres, and $\mathbb{K}(\mathbb{F}) \equiv \mathbb{S}$ the sphere-spectrum.

For the real (respectively, complex) integers $\mathbb{Z}_{\mathbb{R}}$ (respectively, $\mathbb{Z}_{\mathbb{C}}$) we have by Bott periodicity

$$K_n(\mathbb{Z}_{\mathbb{R}}) = \begin{cases} \mathbb{Z} & m \equiv 0 mod 8 \\ \mathbb{Z}/2 & m \equiv 1 mod 8 \\ \mathbb{Z}/2 & n \equiv 2 mod 8 \\ 0 & n \equiv 3 mod 8 \\ \mathbb{Z} & n \equiv 4 mod 8 \\ 0 & n \equiv 5 mod 8 \\ 0 & n \equiv 6 mod 8 \\ 0 & n \equiv 7 mod 8 \end{cases} \tag{16.9}$$

$$K_n(\mathbb{Z}_{\mathbb{C}}) = \begin{cases} \mathbb{Z} \; n \equiv 0 mod 2 \\ 0 \; n \equiv 1 mod 2 \end{cases}$$

For a finite field with q elements, $\mathbb{F}_q$, we have by [**Qui72**]

$$K_n(\mathbb{F}_q) = \begin{cases} \mathbb{Z}\big/(q^i - 1) & n = 2i - 1 \\ 0 & n \text{ even } > 0 \\ \mathbb{Z} & n = 0 \end{cases} \tag{16.10}$$

The truly global spectra associated with $\overline{spec\,\mathbb{Z}}$ is

$$\mathbb{K}(\overline{spec\,\mathbb{Z}}) \equiv \operatorname*{colim}_{\{p_1,\dots,p_k\}\in J} \mathbb{K}(\mathbb{Z}) \prod^{h}_{\mathbb{K}\left(\mathbb{Z}\left[\frac{1}{p_1\cdots p_k}\right]\right)} \mathbb{K}\left(\mathbb{Z}\left[\frac{1}{p_1\dots p_k}\right] \cap \mathbb{Z}_{\mathbb{R}}\right)$$

$$= \mathbb{K}(\mathbb{Z}) \prod^{h}_{\mathbb{K}(\mathbb{Q})} \mathbb{K}(\mathbb{Q} \cap \mathbb{Z}_{\mathbb{R}}) \tag{16.11}$$

or its "real completion":

$$\mathbb{K}(\mathbb{Z}) \prod^{h}_{\mathbb{K}(\mathbb{R})} \mathbb{K}(\mathbb{Z}_{\mathbb{R}})$$

and similarly for number fields.

Note that we have

$$\mathbb{K}(\mathbb{Z}) \prod_{\mathbb{K}(\mathbb{Q})}^{h} \mathbb{K}(\mathbb{Q} \cap \mathbb{Z}_{\mathbb{R}}) \equiv B\mathcal{K}_{\mathbb{Z}} \prod_{B\mathcal{K}_{\mathbb{Q}}}^{h} B\mathcal{K}_{\mathbb{Q}\cap\mathbb{Z}_{\mathbb{R}}} \equiv B\left(\mathcal{K}_{\mathbb{Z}} \overrightarrow{\prod_{\mathcal{K}_{\mathbb{Q}}}} \mathcal{K}_{\mathbb{Q}\cap\mathbb{Z}_{\mathbb{R}}}\right) \tag{16.12}$$

respectively, for its real completion

$$\mathbb{K}(\mathbb{Z}) \prod_{\mathbb{K}(\mathbb{R})}^{h} \mathbb{K}(\mathbb{Z}_{\mathbb{R}}) \equiv B\mathcal{K}_{\mathbb{Z}} \prod_{B\mathcal{K}_{\mathbb{R}}}^{h} B\mathcal{K}_{\mathbb{Z}_{\mathbb{R}}} \equiv B\left(\mathcal{K}_{\mathbb{Z}} \overrightarrow{\prod_{\mathcal{K}_{\mathbb{R}}}} \mathcal{K}_{\mathbb{Z}_{\mathbb{R}}}\right) \tag{16.13}$$

the geometric realization of the comma category of $\mathbb{Z}$-bases that after extension to $\mathbb{Q}$, respectively, $\mathbb{R}$, become orthonormal.

Note that we have a parametrization of all (rational) orthonormal basis via the Cayley transform:

$$\{A \in \mathbb{Q}_{n,m}, A^t = -A\} \xleftrightarrow{\sim} \{g \in \mathrm{SO}_n \cap \mathrm{GL}_n(\mathbb{Q}), \det(I+g) \neq 0\} \tag{16.14}$$

$$\begin{aligned} \{A \in \mathbb{R}_{n,m}, A^t = -A\} &\xleftrightarrow{\sim} \{g \in \mathrm{SO}_n, \det(I+g) \neq 0\} \\ A &\longmapsto (I-A)\circ(I+A)^{-1} \\ (I-g)\circ(I+g)^{-1} &\longleftarrow\!\shortmid\; g \end{aligned} \tag{16.15}$$

and

$$\begin{aligned} O_n \cap \mathrm{GL}_n(\mathbb{Q}) &\equiv (\pm 1)^n \cdot \{g \in \mathrm{SO}_n \cap \mathrm{GL}_n(\mathbb{Q}), \det(I+g) \neq 0\} \\ O_n &\equiv (\pm 1)^n \cdot \{g \in \mathrm{SO}_n, \det(I+g) \neq 0\} \end{aligned} \tag{16.16}$$

Chapter 17

The Witt Ring

For a prop $A \in \mathcal{CProp}$ we have the group $\mathrm{GL}_n(A)$ of invertible elements of the monoid $A_{n,n}$ and it acts on $A_{n,n}$ by conjugation; we let $[p]$ denote the conjugacy class of $p \in A_{n,n}$ and we let $[A_{n,n}]$ denote the collection of these conjugacy classes. We have an embedding

$$[A_{n,n}] \longrightarrow [A_{n+1,n+1}], \quad [p] \longmapsto [p \oplus 0] \tag{17.1}$$

and we take the direct limit:

$$[A] := \lim_{\overrightarrow{n}} [A_{n,n}] \tag{17.2}$$

We have a well defined commutative monoid structure on $[A]$ via

$$[p_1] + [p_2] := [p_1 \oplus p_2] \tag{17.3}$$

Applying the functor K (Grothendieck localization of addition) we get an abelian group, the **Witt group**:

$$\mathcal{W}(A) := K([A]) \tag{17.4}$$

This group has the **Frobenius endomorphisms** given for $n \in \mathbb{N}^+$ by raising square matrices to the nth power

$$F_n : \mathcal{W}(A) \to \mathcal{W}(A), \quad F_n[p] = [p^n] = \Big[\underbrace{p \circ p \circ \cdots \circ p}_{n}\Big],$$

$$F_n \circ F_m = F_{n \cdot m} \tag{17.5}$$

We thus have a functor to Abelian groups with an action of $\mathbb{N}^+$

$$\mathcal{W} : \mathcal{Prop} \longrightarrow (\mathcal{Ab})^{\mathbb{N}^+} \tag{17.6}$$

When the prop A is **totally commutative**, we have multiplication on $[A]$ via, cf. (1.16),

$$[p_1] \cdot [p_2] = [p_1 \otimes p_2] \tag{17.7}$$

It is well defined, commutative and makes $[A]$ into a commutative rig, and so $\mathcal{W}(A)$ is a commutative ring.

The F_n's are **ring endomorphisms**:

$$\mathcal{W} : C_T\mathcal{Prop} \longrightarrow (\mathcal{CRing})^{\mathbb{N}^+} \tag{17.8}$$

We can summarize our constructions in the following diagram:

$$\begin{array}{ccc} \mathcal{Prop} & \xrightarrow{\mathcal{W}} & (\mathcal{Ab})^{\mathbb{N}^+} \\ \cup\!\!| & & \cup\!\!| \\ C_T\mathcal{Prop} & \xrightarrow{\mathcal{W}} & (\mathcal{CRing})^{\mathbb{N}^+} \end{array} \tag{17.9}$$

Here $(\mathcal{CRing})^{\mathbb{N}^+}$ is the category of commutative rings with an action of the multiplicative monoid $\mathbb{N}^+$ of positive natural numbers.

Given a prop-scheme $(\mathcal{X}, \mathcal{O}_\mathcal{X})$ we can apply $\mathcal{W}$ to the props $\mathcal{O}_\mathcal{X}(\mathcal{U})$ to obtain a sheaf of abelian groups $\mathcal{W}(\mathcal{O}_\mathcal{X})$ over $\mathcal{X}$.

When all the $\mathcal{O}_\mathcal{X}(\mathcal{U})$ are totally commutative $\mathcal{W}(\mathcal{O}_\mathcal{X})$ is a sheaf of commutative rings with Frobenius endomorphisms.

Given an ordinary commutative ring $R \in \mathcal{CRing}$, we have the associated totally-commutative prop $R \in C_T\mathcal{Prop}$ and $\mathcal{W}(R)$ is the ring of **rational Witt vectors**, cf. [**Alm74**],

$$\begin{array}{ccc} \mathcal{W}(R) & \hookrightarrow & 1 + x \cdot R[[x]] \\ [p_1] - [p_2] & \longmapsto & \frac{\det(1 - x\cdot p_1)}{\det(1 - x\cdot p_2)} \end{array} \tag{17.10}$$

For a domain R, with fraction field K, let $\overline{K}$ be an algebraic closure of K, $G_K = \mathrm{Gal}\left(\overline{K}/K\right)$ (the absolute Galois group), $\overline{R}$ the integral closure of R in $\overline{K}$, then $\mathcal{W}(R)$ is the free abelian group on monic irreducible polynomials $f(x) \neq x$, via the correspondence

$$f(x) = x^n + a_1 x^{n-1} + \cdots + a_n \longleftrightarrow$$

$$\tilde{f}(x) = x^{\deg f} \cdot f\left(\frac{1}{x}\right) = 1 + a_1 x + \cdots + a_n x^n \tag{17.11}$$

Alternatively, $\mathcal{W}(R)$ is the free abelian group on G_K orbits in $\overline{R}\setminus\{0\}$:

$$\mathcal{W}(R) = \bigoplus_{[\alpha]\in(\overline{R}\setminus\{0\})/G_K} \mathbb{Z}[\alpha] \tag{17.12}$$

with $[\alpha] = \{\alpha = \alpha_1, \ldots, \alpha_n\}$ the G_K-conjugates of α.

The multiplication in $\mathcal{W}(R)$ is given by multiplying orbits:

$$[\alpha]\cdot[\beta] = \{\alpha_1, \ldots, \alpha_n\}\cdot\{\beta_1, \ldots, \beta_m\} = \{\alpha_i \cdot \beta_j\} \tag{17.13}$$

Example 17.1. (The global Witt ring $\mathcal{W} \equiv \mathcal{W}(\overline{\mathit{spec}\ \mathbb{Z}})$). For the compactified $\overline{\mathit{spec}(\mathbb{Z})} = \mathit{spec}(\mathbb{Z}) \cup \{\eta\}$, the condition at the real prime implies that all the eigenvalues of our square matrix are in the unit disc, hence the global sections $\mathcal{W}(\overline{\mathit{spec}(\mathbb{Z})})$ is the free abelian group on $G_{\mathbb{Q}}$-orbits of algebraic integers $[\alpha] = \{\alpha_1, \ldots, \alpha_n\}$ with $|\alpha_i| \leqslant 1$, hence they are **roots of unity**, and so

$$\mathcal{W}(\overline{\mathit{spec}\mathbb{Z}}) \equiv \bigoplus_{n\geqslant 1} \mathbb{Z}\cdot\phi_n = \bigoplus_{n\geqslant 1} \mathbb{Z}[\mu_n^*] \tag{17.14}$$

with ϕ_n the cyclotomic polynomial with roots the primitive nth roots of unity μ_n^*. (Or as the rational functions with zeros and poles in $\mu_\infty \amalg \{\infty\}$:

$$f(x) = \frac{\phi_{n_1}^{e_1}(x)\cdots\phi_{n_k}^{e_k}(x)}{\phi_{m_1}^{d_1}(x)\cdots\phi_{m_\ell}^{d_\ell}(x)}, \qquad e_1, \ldots, e_k, \qquad d_1, \ldots, d_\ell \in \mathbb{N}^+$$

We will not use this multiplicative notations.)

The multiplication is given by $[\mu_n^*]\cdot[\mu_m^*] = [\mu_n^*\cdot\mu_m^*]$, or

$$\phi_n \cdot \phi_m = \phi_{n\cdot m} \quad \text{for} \quad (n, m) = 1 \tag{17.15}$$

$$\phi_{p^n}\cdot\phi_{p^m} = \begin{cases} (1-p^{-1})p^m\cdot\phi_{p^n} & n > m \geqslant 1 \\ (1-p^{-1})p^n\cdot\left[\phi_{p^n} + \phi_{p^{n-1}} + \cdots + \phi_p + \phi_1\right] - p^{n-1}\cdot\phi_{p^n} & n = m \geqslant 1 \end{cases} \tag{17.16}$$

The algebra $\mathcal{W} = \mathcal{W}(\overline{spec\mathbb{Z}})$ is the union of its finite subalgebras

$$\mathcal{W} = \bigcup_{N \geqslant 1} \mathcal{W}_N, \quad \mathcal{W}_N = \bigoplus_{d|N} \mathbb{Z}\phi_d \tag{17.17}$$

It is also the tensor product over all primes (with respect to $\phi_1 = 1$)

$$\mathcal{W} = \otimes_p \mathcal{W}_{p^\infty}, \quad \mathcal{W}_{p^\infty} = \oplus_{\ell \geqslant 0} \mathbb{Z}\phi_{p^\ell} \tag{17.18}$$

with $\mathcal{W}_{p^n}/\mathcal{W}_{p^{n-1}}$ being an integral quadratic extension (17.16). The Frobenius ring endomorphisms are completely multiplicative

$$F_{m_1} \circ F_{m_2} = F_{m_1 \cdot m_2} \tag{17.19}$$

and are given for p prime by

$$F_p\phi_n = \begin{cases} \phi_n & p \nmid n \\ p \cdot \phi_{n/p} & p^2 \mid n \\ (p-1) \cdot \phi_{n/p} & p \mid n, \quad p^2 \nmid n \end{cases} \tag{17.20}$$

or generally by

$$F_m\phi_n = (m,n) \cdot \left(\prod_{\substack{p|(m,n) \\ p \nmid n/(m,n)}} (1 - p^{-1}) \right) \cdot \phi_{n/(m,n)},$$

$$(m,n) = \gcd(m,n) \tag{17.21}$$

We have a cannonical ring homomorphism

$$\begin{aligned} t_1 = \mathrm{tr} : \mathcal{W} &\longrightarrow \mathbb{Z} \\ [a] &\longmapsto \mathrm{tr}\,(a) \end{aligned} \tag{17.22}$$

and

$$\mathrm{tr}(\phi_n) = \sum_{\xi \in \mu_n^*} \xi = \mu(n) \tag{17.23}$$

is the **Möbius function**.

We obtain the homomorphisms

$$\begin{aligned} t_m = \mathrm{tr} \circ F_m : \mathcal{W} &\longrightarrow \mathbb{Z} \\ [a] &\longmapsto \mathrm{tr}(a^m) \end{aligned} \tag{17.24}$$

and

$$t_m(\phi_n) = \mathrm{tr}(F_m \phi_n) = \sum_{\xi \in \mu_n^*} \xi^m = C_n^m \tag{17.25}$$

with the **Ramanujan sums**

$$C_n^m = \mu\left(\frac{n}{(n,m)}\right) \cdot \frac{\varphi(n)}{\varphi(\frac{n}{(n,m)})}, \quad n \geqslant 1, \ \ m \in (\mathbb{Z}/n\mathbb{Z}) \,/\, (\mathbb{Z}/n\mathbb{Z})^* \tag{17.26}$$

Note that

$$m \equiv 0 \pmod n \Longrightarrow F_m \phi_n = \varphi(n) \cdot \phi_1 \tag{17.27}$$

We get by continuity an action of the multiplicative monoid of "super-natural-number"

$$\hat{\mathbb{Z}}/\hat{\mathbb{Z}}^* = \prod_{p \text{ prime}} p^{\mathbb{N} \cup \{\infty\}} \tag{17.28}$$

by ring endomorphisms of $\mathcal{W}$.

In particular, we get the homomorphism

$$F_0 = \lim_{m \to 0 \in \hat{\mathbb{Z}}} F_m \ : \ \mathcal{W} \longrightarrow \mathbb{Z} = \mathbb{Z}\phi_1 \subseteq \mathcal{W} \tag{17.29}$$

We have $F_0 \circ F_m = F_0$ for all m, and

$$F_0 \phi_n = \# \mu_n^* = \varphi(n) = n \prod_{p|n} (1 - p^{-1}) \tag{17.30}$$

is **Euler's function**.

Remark 17.1. The group of roots of unity μ_∞ give rise to the group-ring $\mathbb{Z} \cdot \mu_\infty$, which maps homomorphically onto the ring of cyclotomic integers $\mathbb{Z}[\mu_\infty]$, and also maps onto $\mathbb{Z}$ via the augmutation. Taking $\hat{\mathbb{Z}}^* = \mathrm{Aut}(\mu_\infty)$ invariants we get

$$\mathcal{W} \equiv (\mathbb{Z}\cdot\mu_\infty)^{\hat{\mathbb{Z}}^*} \begin{array}{l} \xrightarrow{t_1} (\mathbb{Z}[\mu_\infty])^{\hat{\mathbb{Z}}^*} = \mathbb{Z} \\ \xrightarrow{F_0} \mathbb{Z} \end{array} \tag{17.31}$$

Remark 17.2. We have

$$\mathcal{CRing}(\mathcal{W},\mathbb{C}) \equiv \mathcal{CRing}(\mathcal{W},\mathbb{Z}) \equiv \hat{\mathbb{Z}}/\hat{\mathbb{Z}}^*$$

Indeed, from (17.18) we obtain

$$\mathcal{CRing}(\mathcal{W},\mathbb{C}) \equiv \prod_p \mathcal{CRing}(\mathcal{W}_{p^\infty},\mathbb{C})$$

From (17.16), we obtain for $\psi \in \mathcal{CRing}(\mathcal{W}_{p^\infty},\mathbb{C})$ the implications for $n \geqslant 1$,

$$\begin{aligned} &\psi(\phi_{p^n}) \neq 0 \\ &\quad\Longrightarrow \psi(\phi_{p^m}) = p^m\left(1-p^{-1}\right) \text{ for } m = 1,2,\ldots,n-1 \\ &\quad\Longrightarrow \psi(\phi_{p^n}) = p^n\left(1-p^{-1}\right) \text{ or } \psi(\phi_{p^n}) = -p^{n-1} \end{aligned}$$

From this it follows that $\psi = \mathrm{tr}\circ F_{p^n}$, $n \geqslant 0$,

$$\mathrm{tr}\circ F_{p^n}(\phi_{p^m}) = \begin{cases} 1 & m = 0 \\ p^m(1-p^{-1}) & 1 \leqslant m \leqslant n \\ -p^n & m = n+1 \\ 0 & m > n+1 \end{cases}$$

or that $\psi = \lim_{n\to\infty} \mathrm{tr}\circ F_{p^n} = F_{p^\infty} = F_0$. Thus

$$\mathcal{CRing}(\mathcal{W}_{p^\infty},\mathbb{C}) \equiv p^{\mathbb{N}\cup\{\infty\}}$$

and note that any $\psi \in \mathcal{CRing}(\mathcal{W},\mathbb{C})$ takes values in $\mathbb{Z}$. □

We have as well the additive projection

$$\int : \mathcal{W} \to \mathbb{Z}, \quad \int \phi_n = \begin{cases} 1 & n = 1 \\ 0 & n > 1 \end{cases} \tag{17.32}$$

and we have

$$\int(\phi_{n_1}\cdot\phi_{n_2}) = \begin{cases}\varphi(n) & n_1 = n_2 = n\\ 0 & n_1 \neq n_2\end{cases} \tag{17.33}$$

We can extend scalar to $\mathbb{C}$, and obtain the ring homomorphisms

$$t_m = \mathrm{tr}\circ F_m : \mathcal{W}_{\mathbb{C}} = \mathbb{C}\otimes\mathcal{W} \longrightarrow \mathbb{C} \tag{17.34}$$

The projection $\int : \mathcal{W}_{\mathbb{C}} \to \mathbb{C}$ can be expanded using the t_m's as

$$\int f \equiv \lim_{M\to 0\in\widehat{\mathbb{Z}}} \frac{1}{M}\sum_{m=1}^{M} t_m(f) \tag{17.35}$$

with the limit taken over a sequence of $M \in \mathbb{N}$ converging to 0 in $\widehat{\mathbb{Z}}$, that is the M's become divisible by anything eventually, e.g.

$$\int f \equiv \lim_{n\to\infty} \frac{1}{n!}\sum_{m=1}^{n!} t_m(f)$$

We can complete $\mathcal{W}_{\mathbb{C}}$ with respect to the state $\int$ and obtain the Hilbert space

$$\mathcal{H} = \widehat{\mathcal{W}}_{\mathbb{C}}, \quad \langle f, g\rangle = \int(f\cdot\overline{g}) \tag{17.36}$$

with orthogonal basis $\{\phi_n\}_{n\geqslant 1}$, $\|\phi_n\|^2 = \varphi(n)$.

The complete-multiplicativity of the F_m, (17.19), suggests considering the **zeta operator**

$$\begin{aligned}\zeta(F,s) &:= \sum_{m\geqslant 1}\frac{1}{m^s}F_m \equiv \prod_p\left(1+\frac{1}{p^s}F_p+\frac{1}{p^{2s}}F_{p^2}+\cdots\right)\\ &\equiv \prod_p \frac{1}{(1-p^{-s}F_p)}\end{aligned} \tag{17.37}$$

The adjoint of F_m, the "verschiebung" F_m^*, is the additive map given by $F_m^*\phi_n = \phi_{m\cdot n}$. The multiplicativity $F_{m_1}^*\circ F_{m_2}^* = F_{m_1\cdot m_2}^*$ suggest considering similarly

$$\zeta(F^*,s) = \sum_{m\geqslant 1}\frac{1}{m^s}F_m^* = \prod_p\frac{1}{(1-p^{-s}F_p^*)} = \zeta(F,\overline{s})^* \tag{17.38}$$

The operators $\zeta(F,s)$ and $\zeta(F^*,s)$ are defined on the dense subspace $\mathcal{W}_{\mathbb{C}} \subseteq \mathcal{H}$ for $\Re(s) > 1$.

In terms of these zeta operators we can interpret Ramanujan's sums:

$$\text{(I) } t_m\big(\zeta(F^*,t)\phi_1\big) = \operatorname{tr}\big(F_m\zeta(F^*,t)\phi_1\big) = \sum_{n\geqslant 1}\frac{C_n^m}{n^t} = \frac{1}{\zeta(t)}\cdot\left(\sum_{d|m} d^{1-t}\right) \tag{17.39}$$

$$\text{(II)} \quad \operatorname{tr}\left(\zeta(F,s)\phi_n\right) = \sum_{m\geqslant 1}\frac{C_n^m}{m^s} = \zeta(s)\cdot\left(\sum_{d|n}\mu\left(\frac{n}{d}\right)d^{1-s}\right) \tag{17.40}$$

$$\text{(III)} \quad \operatorname{tr}\left(\zeta(F,s)\zeta(F^*,t)\phi_1\right) = \sum_{n,m\geqslant 1}\frac{C_n^m}{n^t m^s} = \frac{\zeta(s)}{\zeta(t)}\cdot\zeta(s+t-1) \tag{17.41}$$

with $\zeta(s)$ the Riemann zeta function.

Indeed, we have

$$\sum_{d|n} C_d^m = \sum_{\zeta\in\mu_n}\zeta^m = \begin{cases} n & n|m \\ 0 & n\nmid m\end{cases}$$

and we get by Möbius inversion

$$C_n^m = \sum_{\substack{d|n\\ d|m}}\mu\left(\frac{n}{d}\right)\cdot d \tag{17.42}$$

Therefore,

$$\left(\frac{n}{d} := k\right)$$

$$\text{(I)} \quad \sum_{n\geqslant 1}\frac{C_n^m}{n^t} = \sum_{n\geqslant 1}\sum_{\substack{d|m\\ d|n}}\frac{\mu(\frac{n}{d})\cdot d}{n^t}$$

$$= \sum_{k\geqslant 1}\frac{\mu(k)}{k^t}\cdot\sum_{d|m}\frac{d}{d^t} = \frac{1}{\zeta(t)}\left(\sum_{d|m} d^{1-t}\right)$$

Similarly,

$$\text{(II)} \quad \sum_{m\geqslant 1} \frac{C_n^m}{m^s} = \sum_{m\geqslant 1} \frac{1}{m^s} \sum_{\substack{d|n \\ d|m}} \mu\left(\frac{n}{d}\right) \cdot d = \sum_{d|n} \mu\left(\frac{n}{d}\right) \cdot d \sum_{k\geqslant 1} \frac{1}{(kd)^s}$$

$$= \zeta(s) \left(\sum_{d|n} \mu\left(\frac{n}{d}\right) d^{1-s} \right)$$

and

$$\text{(III)} \quad \sum_{n,m\geqslant 1} \frac{C_n^m}{n^t m^s} = \sum_{n,m\geqslant 1} \frac{1}{n^t m^s} \sum_{\substack{d|n \\ d|m}} \mu\left(\frac{n}{d}\right) \cdot d$$

$$= \sum_{d\geqslant 1} \sum_{\ell,k\geqslant 1} \frac{1}{(\ell d)^t (kd)^s} \mu(\ell) \cdot d$$

$$= \sum_{d\geqslant 1} d^{1-t-s} \sum_{\ell\geqslant 1} \frac{\mu(\ell)}{\ell^t} \sum_{k\geqslant 1} \frac{1}{k^s} = \frac{\zeta(s+t-1)\zeta(s)}{\zeta(t)}$$

The algebra $\mathcal{W} = \mathcal{W}(\overline{\mathit{spec}\,\mathbb{Z}})$, and the Hilbert space $\mathcal{H}$, are truly global objects, but note that the sums in (17.39–17.41) are missing the Γ-factor. We can add these Γ-factors by considering the "compactified zeta operators" which are the Mellin transform of the "theta operator":

$$\hat{\zeta}(F,s) := \zeta_{\mathbb{R}}(s) \cdot \zeta(F,s) \equiv \int_0^\infty d^*x \cdot |x|^s \cdot \left[\sum_{m\geqslant 1} F_m \, e^{-\frac{(mx)^2}{2}} \right]$$

Note that by substituting $\beta := 1-s$, $\alpha := t+s-1$, multiplying (17.41)(III) by the Γ-factors, and using the functional equation (11.38) we obtain for $\Re(\alpha), \Re(\beta) > 1$:

$$\operatorname{tr}\left(\zeta(F,s)\zeta(F^*,t)\phi_1\right) \cdot \frac{\zeta_{\mathbb{R}}(s)\zeta_{\mathbb{R}}(s+t-1)}{\zeta_{\mathbb{R}}(t)} (2\pi)^{s-\frac{1}{2}}$$

$$= \frac{\zeta_{\mathbb{Q}}(s) \cdot \zeta_{\mathbb{Q}}(s+t-1)}{\zeta_{\mathbb{Q}}(t)} \cdot (2\pi)^{s-\frac{1}{2}} = \frac{\zeta_{\mathbb{Q}}(1-s) \cdot \zeta_{\mathbb{Q}}(s+t-1)}{\zeta_{\mathbb{Q}}(t)}$$

$$= \frac{\zeta_{\mathbb{Q}}(\alpha)\cdot\zeta_{\mathbb{Q}}(\beta)}{\zeta_{\mathbb{Q}}(\alpha+\beta)} = \frac{\zeta(\alpha)\cdot\zeta(\beta)}{\zeta(\alpha+\beta)}\cdot\frac{\Gamma\left(\frac{\alpha}{2}\right)\cdot\Gamma\left(\frac{\beta}{2}\right)}{\Gamma\left(\frac{\alpha+\beta}{2}\right)}$$

$$= \left(\sum_{\substack{n,m\geqslant 1\\(n,m)=1}} \frac{1}{n^{\alpha}\cdot m^{\beta}}\right)\cdot\int_{-\infty}^{\infty} d^*x(1+x^2)^{-\alpha/2}(1+x^{-2})^{-\beta/2}$$

$$x := x\cdot m/n$$

$$= \int_{-\infty}^{\infty} d^*x \sum_{\substack{n,m\geqslant 1\\(n,m)=1}} \left(n^2+m^2x^2\right)^{-\frac{\alpha}{2}}\cdot\left(m^2+n^2x^{-2}\right)^{-\frac{\beta}{2}}$$

$$= \int_{-\infty}^{\infty} d^*x\cdot|x|^{\beta}\cdot\sum_{\substack{n,m\geqslant 1\\(n,m)=1}} |n+ixm|^{-(\alpha+\beta)}$$

$$= \int_{0}^{\infty} d^*x\cdot x^{\beta}\cdot\left[\frac{1}{2}\sum_{\substack{n,m\in\mathbb{Z}\\(n,m)=1}} |n+ixm|^{-(\alpha+\beta)} - 1 - x^{-(\alpha+\beta)}\right]$$

$$= \int_{0}^{\infty} d^*x\cdot\left[E(ix,\frac{\alpha+\beta}{2})\cdot x^{\frac{\beta-\alpha}{2}} - x^{\beta} - x^{-\alpha}\right] \tag{17.43}$$

with $E(\tau,\beta)$ the real analytic Eisenstein series, cf. [**Har01**].

The probability measure $\frac{1}{M}\sum_{m=1}^{M}\delta_m$ on $\widehat{\mathbb{Z}}$, converges as $M\to 0\in\widehat{\mathbb{Z}}$, to the Haar measure dz, and we get an isomorphism, the **Fourier transform**

$$\begin{aligned}
\mathcal{H} = \widehat{\mathcal{W}}_{\mathbb{C}} &\xrightarrow{\quad\sim\quad} L_2(\widehat{\mathbb{Z}},dz)^{\widehat{\mathbb{Z}}^*}\\
f &\longmapsto \widehat{f}(z) := \mathrm{tr}(F_z f)\\
F_m &\longmapsto \widehat{F}_m\hat{f}(z) := \hat{f}(mz)\\
F_m^* &\longmapsto \widehat{F}_m^*\hat{f}(z) := m\cdot\hat{f}(z/m)\\
\langle f,g\rangle \quad &= \int_{\widehat{\mathbb{Z}}} \widehat{f}(z)\cdot\overline{\widehat{g}(z)}\,dz
\end{aligned} \tag{17.44}$$

The ring $\mathcal{W}$ has a structure of a special-λ-ring, with λ-operations:

$$\lambda^n : \mathcal{W} \to \mathcal{W}, \quad \lambda^0(x) = 1, \quad \lambda^1(x) = x,$$
$$\lambda^n(x_1 + x_2) = \sum_{n_1+n_2=n} \lambda^{n_1}(x_1) \cdot \lambda^{n_2}(x_2) \tag{17.45}$$

defined by the generating series

$$\begin{array}{ccc} \lambda_t(x) = \sum_{n\geqslant 0}(-1)^n \lambda^n(x) \cdot t^n : \mathcal{W} & \longrightarrow & \left(1 + t \cdot \mathcal{W}[|t|]\right)_{\text{rat}} \\ \text{\rotatebox{90}{$\subseteq$}} & & \text{\rotatebox{90}{$\subseteq$}} \\ \bigoplus_{n\geqslant 0} \mathbb{N} \cdot \phi_n \equiv \mathcal{W}^+ & \longrightarrow & 1 + t \cdot \mathcal{W}[t] \end{array} \tag{17.46}$$

with $\lambda_t(\phi_n)$ given using $\mathcal{W} \equiv (\mathbb{Z} \cdot \mu_\infty)^{\hat{\mathbb{Z}}^*} \subseteq \mathbb{Z} \cdot \mu_\infty$ by the polynomial

$$\lambda_t(\phi_n) = \prod_{\xi \in \mu_n^*} (1 - [\xi]t) = \sum_{m=0}^{\varphi(n)} \lambda^m(\phi_n) \cdot (-t)^m \in \mathcal{W}[t] \tag{17.47}$$

For example,

$$\begin{aligned}
\lambda_t(\phi_1) &= 1 - t \\
\lambda_t(\phi_2) &= 1 - t \cdot \phi_2 \\
\lambda_t(\phi_3) &= 1 - t\phi_3 + t^2 \\
\lambda_t(\phi_4) &= 1 - t\phi_4 + t^2 \\
\lambda_t(\phi_5) &= 1 - t\phi_5 + t^2(\phi_5 + 2\phi_1) - t^3\phi_5 + t^4 \\
\lambda_t(\phi_6) &= 1 - t\phi_6 + t^2 \\
\lambda_t(\phi_7) &= 1 - t\phi_7 + t^2(2 \cdot \phi_7 + 3\phi_1) - t^3(3\phi_7 + 2\phi_1) \\
&\quad + t^4(2\phi_7 + 3\phi_1) - t^5\phi_7 + t^6 \\
\lambda_t(\phi_8) &= 1 - t\phi_8 + t^2(\phi_4 + 2\phi_2 + 2\phi_1) - t^3\phi_8 + t^4 \\
\lambda_t(\phi_9) &= 1 - t\phi_9 + t^2(\phi_9 + 3\phi_3 + 3\phi_1) - t^3(3\phi_9 + \phi_3) \\
&\quad + t^4(\phi_9 + 3\phi_3 + 3\phi_1) - t^5\phi_9 + t^6 \\
\lambda_t(\phi_{10}) &= 1 - t\phi_{10} + t^2(\phi_5 + 2\phi_1) - t^3\phi_{10} + t^4
\end{aligned} \tag{17.48}$$

We have the functional equation

$$\lambda_t(\phi_n) = t^{\varphi(n)} \cdot \lambda_{t^{-1}}(\phi_n) \qquad \text{for } n > 2 \tag{17.49}$$

Applying the homomorphisms of (17.31)

$$t_1 = tr, \quad \text{respectively,} \quad t_0 = F_0 : 1 + t\mathcal{W}[t] \rightrightarrows 1 + t\mathbb{Z}[t]$$

we get

$$t_1\big(\lambda_t(\phi_n)\big) = \prod_{\xi\in\mu_n^*} (1 - \xi \cdot t) = \tilde{\phi}_n(t) = \phi_n(t) \tag{17.50}$$

the cyclotomic polynomial, and respectively,

$$F_0\big(\lambda_t(\phi_n)\big) = (1 - t)^{\varphi(n)} \tag{17.51}$$

Note that $\lambda^n(k \cdot \phi_1) = \binom{k}{n}$ satisfies

$$(-1)^n\lambda^n(-k \cdot \phi_1) = (-1)^n \cdot \binom{-k}{n} = \binom{k+n-1}{n} = \lambda^n(k+n-1) \tag{17.52}$$

We denote the **Grothendieck involution** by

$$D(t) = \begin{bmatrix} 1 & 0 \\ 1 & -1 \end{bmatrix}(t) = \frac{t}{t-1} = \frac{1}{1-t^{-1}} \tag{17.53}$$

so that $DD(t) = t$, $D(0) = 0$, $D(1) = \infty$, $D(\infty) = 1$, $D(2) = 2$, and in the notations of (2.16), (3.27) $1/p + 1/q = 1 \Leftrightarrow q = D(p)$.

We obtain Grothendieck's **gamma-operations** given by the generating series

$$\gamma_t(x) = \sum_{n\geqslant 0} (-1)^n\gamma^n(x)t^n : \mathcal{W} \longrightarrow (1 + t\mathcal{W}[|t|])_{\text{rat}}$$
$$\gamma_t^0(x) = 1, \quad -\gamma^1(x) = x, \quad \gamma^n(x_1 + x_2) = \sum_{n_1+n_2=n} \gamma^{n_1}(x_1)\cdot\gamma^{n_2}(x_2) \tag{17.54}$$

by setting

$$\gamma_t(x) = \lambda_{D(t)}(x) \Leftrightarrow (-1)^n\gamma^n(x) = \lambda^n(x + n - 1), \quad \text{cf. (17.52)} \tag{17.55}$$

We have

$$\mathcal{W} = \mathbb{Z}\cdot\phi_1 \oplus I, \quad I = \ker F_0 = \bigoplus_{n>1} \mathbb{Z}(\phi_n - \varphi(n)\cdot\phi_1) \tag{17.56}$$

The **gamma-filteration** is given by $I_0 = \mathcal{W}$, $I_1 = I$, and for $n \geqslant 2$ by

$$I_n \equiv \text{additive subgroup of } \mathcal{W} \text{ generated by } \gamma^{n_1}(a_1)\cdots\gamma^{n_\ell}(a_\ell),$$
$$a_i \in I, \quad \sum n_i \geqslant n. \quad (17.57)$$

It is a λ-filtration, and on the associated graded λ-ring

$$\mathrm{gr}_\gamma \mathcal{W} \equiv \bigoplus_{n\geqslant 0} I_n/I_{n+1} \quad (17.58)$$

we have

$$F_m \equiv \bigoplus_{n\geqslant 0} m^n \qquad \text{cf. [AT69] Proposition 5.3.} \quad (17.59)$$

and also

$$(-1)^{m+1}\lambda^m \equiv (-1)^{m+1}\binom{x}{m} \oplus \bigoplus_{n\geqslant 1} m^{n-1}$$

cf. [**AT69**] Proposition 5.5. (17.60)

Note that

$$\gamma_t : I^+ \equiv \bigoplus_{n>1} \mathbb{N}\cdot(\phi_n - \varphi(n)\phi_1) \longrightarrow 1 + t\mathcal{W}[t] \quad (17.61)$$

are polynomials and

$$\gamma_t(\phi_n-\varphi(n)\phi_1) = (1-t)^{\varphi(n)}\cdot \prod_{\xi\in\mu_n^*}\left(1-[\xi]\cdot\frac{t}{t-1}\right) = \prod_{\xi\in\mu_n^*}(1-(1-[\xi])t) \quad (17.62)$$

i.e. passing from $\lambda^m(\phi_n)$ to $\gamma^m(\phi_n - \varphi(n)\phi_1)$ amounts to passing from the mth elementary symmetric function of the roots of unity μ_n^*, to that of the "cyclomic units" $1-\xi$, $\xi \in \mu_n^*$, cf. [**Lan78**].

Remark 17.3 (Topological Hochschild Homology as MacLane Homology). Returning to the general case, for any prop A we can define the equivalence relation $\approx$ on the square matrices $\coprod_n A_{n,n}$: $a \approx a'$ iff there exists a path

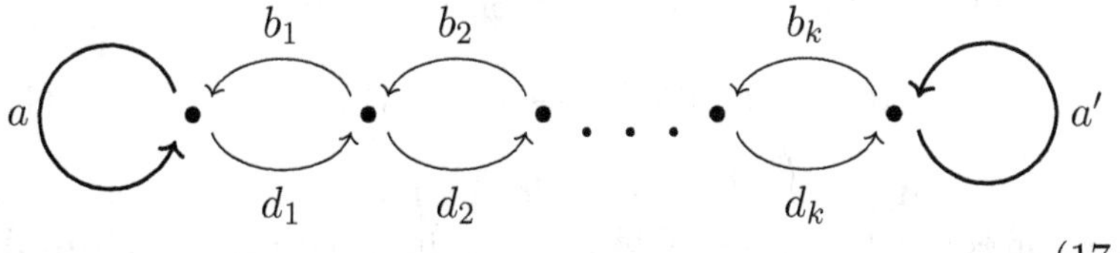

(17.63)

with $a = b_1 \circ d_1$; $d_i \circ b_i = b_{i+1} \circ d_{i+1}, i = 1, \ldots, k-1$; $d_k \circ b_k = a'$. We let $t(a)$ denote the equivalence class of a, and

$$t(A) := \left(\coprod_n A_{n,n} \right) / \approx . \tag{17.64}$$

We have a well defined associative and commutative operation

$$t(a_1) + t(a_2) := t(a_1 \oplus a_2) \tag{17.65}$$

Applying Grothendieck's functor K we get the abelian group

$$\tau(A) := K\left(t(A)\right) \tag{17.66}$$

Note that $\approx$ respects raising a square matrix to an nth power and we get **Frobenius endomorphisms** $F_n : \tau(A) \to \tau(A)$, $F_n t(a) = t(a^n)$.

Note that when A is totally-commutative, so that we have tensor product $a \otimes a'$ of matrices, then $\approx$ respect this operation, $t(A)$ becomes a commutative rig, and $\tau(A)$ is a commutative ring with Frobenius endormorphism

$$\begin{array}{ccc} \mathcal{Prop} & \xrightarrow{\quad \tau \quad} & (\mathcal{Ab})^{\mathbb{N}^+} \\ \cup\!\! | & & \cup\!\! | \\ \mathcal{C}_T\mathcal{Prop} & \xrightarrow{\quad \tau \quad} & (\mathcal{CRing})^{\mathbb{N}^+} \end{array} \tag{17.67}$$

Note that we have

$$a \sim a' \Longrightarrow a \approx a'$$

i.e. if $g \circ a \circ g^{-1} = a'$, then $a = (a \circ g^{-1}) \circ g$, $g \circ (a \circ g^{-1}) = a'$, and $a \approx a'$. Thus we have a well defined map of abelian groups (respectively commutative rings), with Frobenius endomorphisms

$$t : \mathcal{W}(A) \longrightarrow\!\!\!\!\!\rightarrow \tau(A). \tag{17.68}$$

The group $\tau(A)$ is the first **Topological Hochschild Homology** group, or equivalently, cf. [**PW92**], the first **MacLane Homology** group, given by the homology of the cyclic nerve of the category A, $\tau(A) = H_0(\mathbb{Z} \cdot N_{\circlearrowleft}A)$,

$$(N_{\circlearrowleft}A)^* : \coprod_{n_0 \geqslant 0} A_{n_0,n_0} \;\substack{\longleftarrow \\ \dashrightarrow \\ \longleftarrow}\; \coprod_{n_0,n_1 \geqslant 0} A_{n_0,n_1} \times A_{n_1,n_0} \;\substack{\longleftarrow \\ \dashrightarrow \\ \longleftarrow \\ \dashrightarrow \\ \longleftarrow}\; \coprod_{n_0,n_1,n_2 \geqslant 0} A_{n_0,n_1} \times A_{n_1,n_2} \times A_{n_2,n_0} \cdots \tag{17.69}$$

The cyclic nerve is a cyclic set, with a compatible action of $\mathbb{Z}/(d+1) \equiv (S^1)^d$ on $(N_{\circlearrowleft}A)^d$, giving an action of the circle group $|S^1|$ on its geometric realization. The Tate cohomology groups $\widehat{\mathbb{H}}^*(|S^1|, |N_{\circlearrowleft}A|)$ are relevant for zeta functions, cf. [**Hes17**].

The simplicial map

$$\begin{aligned} \mathrm{BGL}_n(A) &\longrightarrow N_{\circlearrowleft}A \\ (g_1, \cdots, g_d) &\rightsquigarrow \left((g_1 \cdot \ldots \cdot g_d)^{-1}, g_1, \ldots, g_d\right) \end{aligned} \tag{17.70}$$

induces the Dennis trace relating the K-groups of A and the Topological Cyclic Homology of $N_{\circlearrowleft}A$, cf. [**DM94**], [**Mad95**].

We end with some remarks on "stable" limits. Note that in (17.69), taking all the n_i's to be $\geqslant n$, we get a sub-cyclic-set, $(N_{\circlearrowleft}A)_{\geqslant n}$, and we have mapping of cyclic sets

$$\begin{aligned} (N_{\circlearrowleft}A)_{\geqslant n} &\rightleftarrows (N_{\circlearrowleft}A)_{\geqslant n+1} \\ (f_0, \ldots, f_d) &\longmapsto (f_0 \oplus (1), \ldots, f_d \oplus (1)) \end{aligned} \tag{17.71}$$

We can take the direct limit with respect to $f_\bullet \longmapsto f_\bullet \oplus (1)$,

$$(N_{\circlearrowleft}A)_\infty := \lim_{n \to \infty} (N_{\circlearrowleft}A)_{\geqslant n} \tag{17.72}$$

Note that the limit with respect to $f_0 \mapsto f_0 \oplus (1)$,

$$A_{\{n_0 - n_1\}} := \lim_{n \to \infty} A_{n_0+n, n_1+n} \qquad \text{depends only on } n_0 - n_1 \in \mathbb{Z}. \tag{17.73}$$

Putting formally (with t a central variable)

$$A\{t\} := \coprod_{n \in \mathbb{Z}} A_{\{n\}} \cdot t^n \tag{17.74}$$

we see that $(N_{\circlearrowleft} A)_\infty$ is the coefficient of t^0 in the Cyclic set

$$A\{t\} \quad \begin{array}{c}\longleftarrow \\ \dashrightarrow \\ \longleftarrow\end{array} A\{t\} \times A\{t\} \begin{array}{c}\longleftarrow \\ \dashrightarrow \\ \longleftarrow \\ \dashrightarrow \\ \longleftarrow\end{array} A\{t\} \times A\{t\} \times A\{t\} \quad \cdots \tag{17.75}$$

The map (17.70) take values in the sub-cyclic-set of "square matrices"

$$A_{\{0\}} \begin{array}{c}\longleftarrow \\ \dashrightarrow \\ \longleftarrow\end{array} A_{\{0\}} \times A_{\{0\}} \begin{array}{c}\longleftarrow \\ \dashrightarrow \\ \longleftarrow \\ \dashrightarrow \\ \longleftarrow\end{array} A_{\{0\}} \times A_{\{0\}} \times A_{\{0\}} \quad \cdots \tag{17.76}$$

Chapter 18

Modules Over the Sphere Spectrum

Let $A \in \mathcal{CBio}$, (e.g. $A = UP$, $P \in \mathcal{CProp}$ and A can be also a simplicial bio), we will study the symmetric-monoidal structure on $A^1[\mathcal{Set}]$. Everything works quite generally, but in order for this symmetric-monoidal structure to be **closed**, that is to have a well defined inner Hom functor, we will need to assume our A-sets are **totally-commutative**, so we write for short

$$A\text{-}\mathcal{Set} \equiv C_T A^1[\mathcal{Set}] \tag{18.1}$$

Thus an A-set is a pointed set $\mathfrak{a}$ together with an A-action

$$\begin{aligned} A^-(n) \times \mathfrak{a}^n \times A^+(n) &\longrightarrow \mathfrak{a} \\ b, x_1, \ldots, x_n, d &\rightsquigarrow b \circ (x_i) \circ d \equiv \langle b, x_i, d \rangle \end{aligned} \tag{18.2}$$

satisfying,

Zero:	$\langle 0, x_i, d \rangle \equiv 0 \equiv \langle b, x_i, 0 \rangle$
S_n-covariant:	$\langle b \circ \sigma, x_i, \sigma^{-1} \circ d \rangle \equiv \langle b, x_{\sigma(i)}, d \rangle \quad \sigma \in S_n.$
Associativity:	$\langle b \circ (b_i), x_{i,j}, (d_i) \circ d \rangle \equiv \langle b, \langle b_i, x_{i,j}, d_i \rangle, d \rangle$
Unit:	$\langle 1, x, 1 \rangle \equiv x$

Full-commutative: For $b \in A_{1,\sum_{i=1}^k m_i}$, $d \in A_{k,1}$ and $d_i \in A_{m_i,1}$ we have:

$$\left\langle b, \underbrace{x_1, \ldots, x_1}_{m_1} \cdots \underbrace{x_k \cdots x_k}_{m_k}, (d_1 \cdots d_k) \circ d \right\rangle \equiv \langle b \overleftarrow{\circ} (d_1 \cdots d_k), x_i, d \rangle$$

and similarly for $b \in A_{1,k}$, $b_i \in A_{1,m_i}$, and $d \in A_{\sum_{i=1}^{k} m_i,1}$ we have:

$$\left\langle b \circ (b_1 \cdots b_k), \underbrace{x_1 \cdots x_1}_{m_1} \cdots \underbrace{x_k \cdots x_k}_{m_k}, d \right\rangle \equiv \langle b, x_i, (b_1, \ldots, b_k) \vec{\circ}\, d \rangle$$

and

Totally-commutative: For $b \in A_{1,n}$, $d \in A_{n,1}$, $b' \in A_{1,n'}$, $d' \in A_{n',1}$, we have:

$$\langle b \circ (\underbrace{b', \ldots, b'}_{n}), x_i, (\underbrace{d', \ldots, d'}_{n}) \circ d \rangle$$

$$\equiv \langle b' \circ (\underbrace{b, \ldots, b}_{n'}), x_{\sigma_{n',n}(i)}, (\underbrace{d, \ldots, d}_{n'}) \circ d' \rangle$$

We will not use full-commutativity, but total-commutativity will be essential.

For A-sets M, N, K let the "bilinear-maps" be defined by

$$\begin{aligned} &\mathrm{Bil}_A(M, N; K) \\ &= \left\{ \varphi : M \wedge N \to K, \quad \begin{matrix} \varphi(\langle b, m_j, d \rangle, n) = \langle b, \varphi(m_j, n), d \rangle \\ \varphi(m, \langle b, n_j, d \rangle) = \langle b, \varphi(m, n_j), d \rangle \end{matrix} \right\} \end{aligned} \tag{18.3}$$

It is a functor in K, and is representable

$$\mathrm{Bil}_A(M, N; K) \equiv A\text{-}\mathcal{S}et(M \otimes_A N, K) \tag{18.4}$$

where $M \otimes_A N$ is the totally-commutative A-$\mathcal{S}et$ freely generated by $M \wedge N$ modulo the equivalence relation generated by the A-bilinear relations, and

$$\begin{aligned} M \wedge N &\longrightarrow M \otimes_A N \\ (m, n) &\rightsquigarrow m \otimes n \end{aligned} \tag{18.5}$$

is the universal bilinear map. We thus get a bi-functor

$$\_\otimes_A\_ : A\text{-}\mathcal{S}et \times A\text{-}\mathcal{S}et \longrightarrow A\text{-}\mathcal{S}et \tag{18.6}$$

giving a symmetric monoidal structure on A-$\mathcal{S}et$, with unit

$A(1)^T$ (= the totally-commutative quotient of the A-$\mathcal{S}et$ $A(1)$)

Having assumed that our A-$\mathcal{S}ets$ are totally commutative we have the inner Hom functor

$$\begin{gathered}\mathrm{Hom}_A(\_,\_) : (A\text{-}\mathcal{S}et)^{\mathrm{op}} \times A\text{-}\mathcal{S}et \longrightarrow A\text{-}\mathcal{S}et \\ \mathrm{Hom}_A(M, N) = A\text{-}\mathcal{S}et(M, N)\end{gathered} \tag{18.7}$$

with the A-action given by

$$\begin{aligned}&\langle b, \varphi_j, d\rangle(m) = \langle b, \varphi_j(m), d\rangle, \\ &b \in A^-(n), d \in A^+(n), \varphi_j \in \mathrm{Hom}_A(M, N)\end{aligned} \tag{18.8}$$

This in itself is a map of A-$\mathcal{S}et$ because of total commutativity, and all the other properties (associating, unit, full and total-commutativity) follow from their validity in N.

We have the adjunction

$$A\text{-}\mathcal{S}et(M \otimes_A N, K) = A\text{-}\mathcal{S}et(M, \mathrm{Hom}_A(N, K))$$

For a homomorphism of bios $\varphi : B \to A$, we have the adjunction

$$\begin{array}{c} A\text{-}\mathcal{S}et \\ \varphi_* \uparrow \quad \downarrow \varphi^* \\ B\text{-}\mathcal{S}et \end{array} \tag{18.9}$$

and

$$\varphi_*(M \otimes_B N) = \varphi_*(M) \otimes_A \varphi_*(N), \quad \varphi_* B(1)^T = A(1)^T \tag{18.10}$$

We have therefore the adjunction formula

$$\varphi^* \mathrm{Hom}_A(\varphi_* M, N) \cong \mathrm{Hom}_B(M, \varphi^* N). \tag{18.11}$$

The tensor product is distributive over sums

$$M \otimes_A \left(\coprod_i N_i \right) \cong \coprod_i (M \otimes_A N_i) \tag{18.12}$$

and more generally is distributive over co-limits.

The category $\mathcal{S}A\text{-}\mathcal{S}et \equiv (A\text{-}\mathcal{S}et)^{\mathbb{\Delta}^{\mathrm{op}}}$ of simlicial A-Sets inherit a closed symmetric monoidal structure

$$\begin{aligned}
&-\otimes_A- : \mathcal{S}A\text{-}\mathcal{S}et \times \mathcal{S}A\text{-}\mathcal{S}et \longrightarrow \mathcal{S}A\text{-}\mathcal{S}et,\\
&(M \otimes_A N)^n := M^n \otimes_A N^n\\
&\mathrm{Hom}_A(-,\ -) : (\mathcal{S}A\text{-}\mathcal{S}et)^{\mathrm{op}} \times \mathcal{S}A\text{-}\mathcal{S}et \longrightarrow \mathcal{S}A\text{-}\mathcal{S}et\\
&\mathrm{Hom}_A(M,N)^n := \mathcal{S}A\text{-}\mathcal{S}et(M \otimes \phi_{A*}\Delta(n)_+, N)
\end{aligned} \tag{18.13}$$

Here the cannonical homomorphism

$$\phi_A : \mathbb{F} \longrightarrow A \tag{18.14}$$

gives the adjunction

$$\begin{array}{c}
\mathcal{S}A\text{-}\mathcal{S}et\\
\phi_{A*} \uparrow \quad \downarrow \phi_A^*\\
\mathcal{S}\mathbb{F}\text{-}\mathcal{S}et \equiv \mathcal{S}\mathcal{S}et_0
\end{array} \tag{18.15}$$

and $\phi_{A*}\Delta(n)_+$ is the free totally commutative simplicial $A\text{-}\mathcal{S}et$ on $\Delta(n)$: $\mathcal{S}A\text{-}\mathcal{S}et(\phi_{A*}\Delta(n)_+, N) = \mathcal{S}\mathcal{S}et(\Delta(n), N) = N^n$.

The projective model structure on $\mathcal{S}A\text{-}\mathcal{S}et$, is compatible with the symmetric monoidal structure, but not having addition — it is not stable. We describe its stabilization using the symmetric spectra of [**Hov01**].

We have the category

$$\mathrm{GL}(\mathbb{F}) \equiv \coprod_{n\geqslant 0} \mathrm{GL}_n(\mathbb{F}) \equiv \coprod_{n\geqslant 0} S_n \cong \mathrm{Iso}(\mathrm{FinSet}) \tag{18.16}$$

which is equivalent to the category of bijections of Finite Sets.

The functor category

$$\Sigma(A) := (\mathcal{S}A\text{-}\mathcal{S}et)^{\mathrm{GL}(\mathbb{F})} \cong (\mathcal{S}A\text{-}\mathcal{S}et)^{\mathrm{Iso(FinSet)}} \tag{18.17}$$

has objects the **symmetric-sequences** $M = \{M^n\}_{n\geqslant 0}$, with $M^n \in (\mathcal{S}A\text{-}\mathcal{S}et)^{S_n}$ a simplicial $A\text{-}\mathcal{S}et$ with an action of S_n (so it has a set of d-simplices $M^{n,d}$).

The category $\Sigma(A)$ is complete and co-complete.
It has a closed symmetric monoidal structure,

$$
\begin{aligned}
\_\otimes_{\Sigma(A)}\_ &: \Sigma(A) \times \Sigma(A) \longrightarrow \Sigma(A) \\
(M \otimes_{\Sigma(A)} N)^n &:= \coprod_{p+q=n} S_n \underset{S_p \times S_q}{\times} (M^p \otimes_A N^q)
\end{aligned}
\tag{18.18}
$$

Here the induction functor is the left adjoint of the forgetful functor

$$(\mathcal{S}A\text{-}\mathcal{S}et)^{S_n} \longrightarrow (\mathcal{S}A\text{-}\mathcal{S}et)^{S_p \times S_q}$$

and is given by

$$S_n \underset{S_p \times S_q}{\times} (M) := \coprod_{S_n/S_p \times S_q} M \tag{18.19}$$

Equivalently, writing $M, N \in \Sigma(A)$ as functors
$\mathrm{Iso(FinSet)} \to \mathcal{S}A\text{-}\mathcal{S}et$ we have

$$(M \otimes_{\Sigma(A)} N)^n = \coprod_{n = n_0 \amalg n_1} M^{n_0} \otimes_A N^{n_1} \tag{18.20}$$

the sum over all decompositions of n as a disjoint union of subsets $n_0, n_1 \subseteq n$.

The unit of this monoidal structure is the symmetric sequence

$$\mathbb{1}_A := (A(1)^T, 0, 0, 0, \ldots) \tag{18.21}$$

Note that this monoidal structure is symmetric

$$\mathcal{Y}_{M,N} : M \otimes_{\Sigma(A)} N \cong N \otimes_{\Sigma(A)} M \tag{18.22}$$

This symmetry is clear in the formula (18.20),

$$M^{n_0} \otimes_A N^{n_1} \cong N^{n_1} \otimes_A M^{n_0}$$

but in the formulation of formula (18.18) the symmetry isomorphism

$$S_n \underset{S_p \times S_q}{\times} (M^p \otimes_A N^q) \cong S_n \underset{S_q \times S_p}{\times} (N^q \otimes_A M^p)$$

involves the (p, q)-shuffle $\tau_{p,q} \in S_n$ that conjugates $S_p \times S_q$ to $S_q \times S_p$.

The internal Hom is given by

$$
\begin{aligned}
&\mathrm{Hom}_{\Sigma(A)}(\_,\_) : \Sigma(A)^{\mathrm{op}} \times \Sigma(A) \longrightarrow \Sigma(A) \\
&\mathrm{Hom}_{\Sigma(A)}(M,N)^n := \textstyle\prod_{k \geqslant 0} \mathrm{Hom}_A(M^k, N^{k+n})
\end{aligned}
\tag{18.23}
$$

$$
\Sigma(A)(M \otimes_{\Sigma(A)} N, K) \equiv \Sigma(A)(M, \mathrm{Hom}_{\Sigma(A)}(N,K)) \tag{18.24}
$$

For a homomorphism $\varphi : B \to A$ we have adjunction

$$
\Sigma(B) \underset{\varphi^*}{\overset{\varphi_*}{\rightleftarrows}} \Sigma(A) \tag{18.25}
$$

and φ_* is monoidal

$$
\varphi_*(M \otimes_{\Sigma(B)} N) \cong \varphi_*(M) \otimes_{\Sigma(A)} \varphi_*(N), \quad \varphi_*(\mathbb{1}_B) \cong \mathbb{1}_A \tag{18.26}
$$

Remark 18.1. One can think of the elements of $\Sigma(A)$ as "Fourier-coefficients", and associate with $M \in \Sigma(A)$ the "**Fourier-transform**" given by the "**analytic functor**"

$$
\begin{aligned}
&\widehat{M} : \mathscr{S}A\text{-}\mathcal{S}et \longrightarrow \mathscr{S}A\text{-}\mathcal{S}et \\
&\widehat{M}(X) := \coprod_{n \geqslant 0} M^n \otimes_A^{S_n} X^{\prod n} \\
&M^n \otimes_A^{S_n} X^{\prod n} := \\
&M^n \otimes_A (\underbrace{X \textstyle\prod \cdots \prod X}_{n}) \big/ \sigma m \otimes (x_1, \ldots, x_n) \sim m \otimes (x_{\sigma(1)}, \ldots, x_{\sigma(n)})
\end{aligned}
\tag{18.27}
$$

The Fourier transform converts "convolution" ≡ the symmetric product in $\Sigma(A)$, into "multiplication":

$$\begin{aligned}(M \widehat{\otimes_{\Sigma(A)}} N)(X) &= \coprod_{n\geqslant 0} (M \otimes_{\Sigma(A)} N)^n \otimes_A^{S_n} X^{\prod n} \\ &= \coprod_{p,q\geqslant 0} (M^p \otimes_A N^q) \underset{S_p\times S_q}{\times} S_{p+q} \otimes_A^{S_{p+q}} X^{\prod(p+q)} \\ &= \left(\coprod_{p\geqslant 0} M^p \otimes_A^{S_p} X^{\prod p}\right) \otimes_A \left(\coprod_{q\geqslant 0} N^q \otimes_A^{S_q} X^{\prod q}\right) \\ &= \widehat{M}(X) \otimes_A \hat{N}(X). \end{aligned} \tag{18.28}$$

The composition of analytic functors is again analytic,

$$\begin{aligned} &\widehat{M} \circ \hat{N}(X) := \widehat{M}(\hat{N}(X)) \equiv (\widehat{M @N})(X) \\ &(M @N)^n = \coprod_{m\geqslant 0} \coprod_{\substack{k_1,\cdots,k_m\geqslant 0 \\ k_1+\cdots k_m=n}} M^m \underset{A}{\otimes} (N^{k_1} \prod \cdots \prod N^{k_m}) \end{aligned} \tag{18.29}$$

This gives another (non-symmetric) monoidal structure on $\Sigma(A)$, with unit $\delta = (0, A(1)^T, 0, \ldots)$.

Example 18.1. The category of (simplicial) symmetric operads is precisely the category of monoid objects of $\Sigma(\mathbb{F})$

$$\begin{aligned} &\mathscr{S}\text{Operad} \equiv \text{Mon}(\Sigma(\mathbb{F}), @) \\ &\Cup \\ &M \Longleftrightarrow m : M @M \to M \text{ associative and unital} \end{aligned} \tag{18.30}$$

Note that we can replace "Simplical-Sets" everywhere by Topological spaces, and form the category $\sum_{\mathcal{T}op}(A)$, and we have the realization-functor

$$\Sigma(A) \to \Sigma_{\mathcal{T}op}(A), \quad X \mapsto |X| \tag{18.31}$$

Now the **topological symmetric operads** are the monoids of $\Sigma_{\mathcal{T}op}(\mathbb{F})$

$$\text{Operad}(\mathcal{T}op) \equiv \text{Mon}\left(\Sigma_{\mathcal{T}op}(\mathbb{F}), @\right) \tag{18.32}$$

Example 18.2. We have cf. (11.24) with $K = \mathbb{R}$, the **Sphere Operad**

$$\mathbb{S}_{\mathbb{R}}^{(-)} = (0, S^0 = \{\pm 1\}, S^1, \ldots, S^{n-1}, \ldots) \in \text{Mon}\left(\Sigma_{\mathcal{T}op}(\mathbb{F}), @\right) \tag{18.33}$$

with the $(n-1)$-dimensional sphere in dimension n, with the natural action of S_n on $S^{n-1} \subseteq \mathbb{R}^n$, fixing $\pm\frac{1}{\sqrt{n}}(1, 1, \ldots, 1)$.

The category

$$\Sigma(\mathbb{F}) \equiv (\mathcal{S}\mathcal{S}et_0)^{\text{GL}(\mathbb{F})} \equiv (\mathcal{S}et_0)^{\text{GL}(\mathbb{F}) \times \Delta^{\text{op}}} \tag{18.34}$$

is the usual category of symmetric-sequences of pointed-simplicial-sets, and it contains the **sphere-spectrum**

$$S_{\mathbb{F}}^{\cdot} := \Big\{ S^n = \underbrace{S^1 \wedge \cdot \wedge S^1}_{n} \Big\}_{n \geqslant 0} \tag{18.35}$$

with the permutation action of S_n on S^n.

The sphere-specturm is a monoid object of $\Sigma(\mathbb{F})$, with respect to the symmetric monoidal structure, with multiplication

$$\begin{aligned} &m : S_{\mathbb{F}}^{\cdot} \otimes_{\Sigma(\mathbb{F})} S_{\mathbb{F}}^{\cdot} \longrightarrow S_{\mathbb{F}}^{\cdot} \\ &m(S^n \otimes_{\mathbb{F}} S^m) \equiv m(S^n \wedge S^m) \equiv S^n \wedge S^m \equiv S^{n+m} \end{aligned} \tag{18.36}$$

and it is a **commutative** monoid, $m = m \circ \mathcal{Y}_{S_{\mathbb{F}}^{\cdot}, S_{\mathbb{F}}^{\cdot}}$, cf. (18.20).

The unit of this monoid is given by

$$\varepsilon : \mathbb{1}_{\mathbb{F}} = (\mathbb{F}(1), 0, 0, \ldots) \equiv (S^0, 0, 0, \ldots) \subseteq (S^0, S^1, S^2, \ldots) \equiv S_{\mathbb{F}}^{\cdot} \tag{18.37}$$

using the identification

$$\mathbb{F}(1) \equiv \{0, 1\} \equiv S^0 \tag{18.38}$$

Using the canonical map $\phi_A : \mathbb{F} \to A$, we get the commutative monoid

$$S_A^{\cdot} := \phi_{A*} S_{\mathbb{F}}^{\cdot} \in \Sigma(A) \tag{18.39}$$

We get the category of $S^{\cdot}_A$-modules, $S^{\cdot}_A\text{-}\mathit{mod}$, with objects the symmetric sequences $M^{\cdot} = \{M^n\} \in \Sigma(A)$, together with associative unital $S^{\cdot}_A$-action

$$\mathfrak{a}_M : S^{\cdot}_A \otimes_{\Sigma(A)} M^{\cdot} \longrightarrow M^{\cdot} \tag{18.40}$$

or equivalently, associative, unital, $S_p \times S_q \subseteq S_{p+q}$ covariant, action

$$S^p \wedge M^q \longrightarrow M^{p+q} \tag{18.41}$$

The maps in $S^{\cdot}_A\text{-}\mathit{mod}$ are the maps of $\Sigma(A)$ that preserves the S_A-action.

The category $S^{\cdot}_A\text{-}\mathit{mod}$ is complete and co-complete.

It is a closed symmetric monoidal category with unit $S^{\cdot}_A$, with respect to

$$\_\otimes_{S^{\cdot}_A}\_ : S^{\cdot}_A\text{-}\mathit{mod} \times S_A\text{-}\mathit{mod} \longrightarrow S^{\cdot}_A\text{-}\mathit{mod}$$
$$M^{\cdot} \otimes_{S^{\cdot}_A} N^{\cdot}$$
$$:= \operatorname{colim}\left\{ M^{\cdot} \otimes_{\Sigma(A)} S^{\cdot}_A \otimes_{\Sigma(A)} N^{\cdot} \underset{\mathrm{Id}_M \otimes a_N}{\overset{\mathfrak{a}_M \otimes \mathrm{Id}_N}{\rightrightarrows}} M^{\cdot} \otimes_{\Sigma(A)} N^{\cdot} \right\}$$
$$\mathrm{Hom}_{S^{\cdot}_A}(\_,\_) : (S^{\cdot}_A\text{-}\mathit{mod})^{\mathrm{op}} \times S^{\cdot}_A\text{-}\mathit{mod} \longrightarrow S^{\cdot}_A\text{-}\mathit{mod}$$
$$\mathrm{Hom}_{S^{\cdot}_A}(M^{\cdot}, N^{\cdot})$$
$$:= \lim\left\{ \mathrm{Hom}_{\Sigma(A)}(M^{\cdot}, N^{\cdot}) \rightrightarrows \mathrm{Hom}_{\Sigma(A)}(S^{\cdot}_A \otimes_{\Sigma(A)} M^{\cdot}, N^{\cdot}) \right\} \tag{18.42}$$

For $M^{\cdot}, N^{\cdot} \in S^{\cdot}_A\text{-}\mathit{mod}$, we have also the simplicial space

$$\mathrm{Map}_{S^{\cdot}_A}(M^{\cdot}, N^{\cdot})^n := S^{\cdot}_A\text{-}\mathit{mod}(M^{\cdot} \otimes \phi_{A*}\Delta(n)_+, N^{\cdot}) \tag{18.43}$$

We have the projective model structure on $S^{\cdot}_A\text{-}\mathit{mod}$ where the fibrations and weak-equivalences are defined "level wise"

$$\begin{aligned} \mathcal{F}ib^{\mathrm{lev}} &= \{f^{\cdot} \in S^{\cdot}_A\text{-}\mathit{mod}(M^{\cdot}, N^{\cdot}), f^n \in \mathcal{F}ib_A \text{ all } n \geqslant 0\} \\ \mathcal{W}^{\mathrm{lev}} &= \{f^{\cdot} \in S^{\cdot}_A\text{-}\mathit{mod}(M^{\cdot}, N^{\cdot}), f^n \in \mathcal{W}_A \text{ all } n \geqslant 0\} \\ \mathcal{C}of_{S^{\cdot}_A} &= \mathcal{L}\{\mathcal{W}^{\mathrm{lev}} \cap \mathcal{F}ib^{\mathrm{lev}}\} \end{aligned} \tag{18.44}$$

We let $\Omega_A\text{-}\mathit{mod} \subseteq S^{\cdot}_A\text{-}\mathit{mod}$ denote the full subcategory of Ω-spectrum, consisting of the $M^{\cdot} = \{M^n\} \in S^{\cdot}_A\text{-}\mathit{mod}$ which are levelwise fibrant, and the adjoint of the action map

$$a_M^{1,n} : S_A^1 \otimes_A M^n \longrightarrow M^{n+1}$$

is a weak-equivalence

$$(a_M^{1,n})^{\#} : M^n \xrightarrow{\sim} \mathrm{Hom}_A(S_A^1, M^{n+1}) =: \Omega M^{n+1} \tag{18.45}$$

The **stable** model structure on $S^{\cdot}_A$-$\mathcal{mod}$ is a Boushfield localization of the projective model structure, having the same cofibrations $\mathcal{Cof}_{S^{\cdot}_A}$, but with the fibrant objects being the Ω-spectra Ω_A-$\mathcal{mod}$:

$$\mathcal{W}_{S^{\cdot}_A} = \left\{ \begin{array}{l} f^{\cdot} \in S^{\cdot}_A\text{-}\mathcal{mod}(M^{\cdot}, N^{\cdot}),\ \mathrm{Map}(N^{\cdot}, X^{\cdot}) \xrightarrow{\sim} \mathrm{Map}(M^{\cdot}, X^{\cdot}) \text{ is a} \\ \qquad\qquad\qquad\qquad\qquad\qquad \text{weak-equivalence of simplicial sets} \\ \qquad\qquad\qquad\qquad\qquad\qquad \text{for all } X^{\cdot} \in \Omega_A\text{-}\mathcal{mod} \end{array} \right\}$$
$$\mathcal{Fib}_{S^{\cdot}_A} = \mathcal{R}\{\mathcal{Cof}_{S^{\cdot}_A} \cap \mathcal{W}_{S^{\cdot}_A}\} \tag{18.46}$$

We have the Quillen adjunctions, inducing a Quillen equivalence,

$$\_\otimes_A S_A^1 \equiv \_\otimes_{\mathbb{F}} S^1 : S^{\cdot}_A\text{-}\mathcal{mod} \rightleftarrows S^{\cdot}_A\text{-}\mathcal{mod} : \mathrm{Hom}_A(S_A^1, \_) \equiv \Omega(\_) \tag{18.47}$$

The **derived** category of $A^{\cdot}$-$\mathcal{Set}$, is the associated homotopy category

$$\mathbb{D}(A\text{-}\mathcal{Set}) := \mathrm{Ho}(S^{\cdot}_A\text{-}\mathcal{mod}) \cong S^{\cdot}_A\text{-}\mathcal{mod}\left[\mathcal{W}_{S^{\cdot}_A}^{-1}\right] \tag{18.48}$$

It is a triangulated, closed symmetric monoidal category (with respect to the derived functors $\_ \otimes^{\mathbb{L}}_{S^{\cdot}_A} \_$ and $\mathbb{R}\mathrm{Hom}_{S^{\cdot}_A}(\_,\_)$).

Given a homomorphism $\varphi : B \to A$ we get the adjunction

$$\mathbb{D}(A\text{-}\mathcal{S}et) \underset{\mathbb{R}\varphi^*}{\overset{\mathbb{L}\varphi_*}{\leftrightarrows}} \mathbb{D}(B\text{-}\mathcal{S}et) \tag{18.49}$$

and $\mathbb{L}\varphi_*$ is a monoidal functor commuting with $\_\otimes^{\mathbb{L}}_A S^1_A$ and $\mathbb{R}\Omega(\_)$.

All this generalizes to a global prop or bio scheme X by taking the (infinite categorical) limit of the fibrant-cofibrant S_A-modules taken over all affine open $\mathit{spec}(A) \subseteq X$. By using Lurie's (un)straightening functors the higher-coherencies conditions describing the derived category $\mathbb{D}(X)$ of quasi-coherent $\mathcal{O}_X$-$\mathcal{S}et$ can be spelled out explicitly, see [**Har20**]. For a quasi-coherent map of generalized schemes $f : X \to Y$ (15.10) (vii) we get adjunction as in (18.49).

Passing to pro-generalized schemes, as we did in chapter 10, things get even simplified, (cf., e.g. [**BS16**]).

Bibliography

[Alm74] Gert Almkvist. The Grothendieck ring of the category of endomorphisms. *J. Algebra*, 28(3):375–388, 1974.

[AM69] Mike Artin and Barry Mazur. *Etale homotopy*. Lecture Notes in Mathematics, No. 100. Springer-Verlag, Berlin-New York, 1969.

[Ara74] S. Ju Arakelov. Intersection theory of divisors on an arithmetic surface. *Mathematics of the USSR-Izvestiya*, 8(6):1167, 1974.

[Art06] Emil Artin. *Algebraic Numbers and Algebraic Functions*. AMS Chelsea Publishing, Providence, RI, 2006. Reprint of the 1967 original.

[AT69] M. F. Atiyah and D. O. Tall. Group representations, λ-rings and the J-homomorphism. *Topology*, 8:253–297, 1969.

[BG72] Kenneth S. Brown and Stephen M. Gersten. Algebraic K-theory as generalized sheaf cohomology. In *Higher K-Theories: Proceedings of the Conference held at the Seattle Research Center of the Battelle Memorial Institute, from August 28 to September 8, 1972*, pp. 266–292. Springer, 1972.

[BL96] Jonathan Block and Andrey Lazarev. Homotopy theory and generalized duality for spectral sheaves. *Internat. Math. Res. Notices*, (20):983–996, 1996.

[Bor09] James Borger. Lambda-rings and the field with one element. *arXiv preprint arXiv:0906.3146*, 2009.

[BP72] Michael Barratt and Stewart Priddy. On the homology of non-connected monoids and their associated groups. *Comment. Math. Helv.*, 47:1–14, 1972.

[BS16] Ilan Barnea and Tomer M. Schlank. A projective model structure on pro-simplicial sheaves, and the relative étale homotopy type. *Adv. Math.*, 291:784–858, 2016.

[BV73] J. M. Boardman and R. M. Vogt. *Homotopy Invariant Algebraic Structures on Topological Spaces*, volume 347 of *Lecture Notes in Mathematics.* Springer-Verlag, Berlin-New York, 1973.

[CC16] Alain Connes and Caterina Consani. Absolute algebra and Segal's gamma sets. *J. Number Theory*, 162:518–551, 2016.

[CF67] J. W. S. Cassels and A. Fröhlich, editors. *Algebraic Number Theory: Proceedings of an Instructional Conference Organized by the London Mathematical Society (a NATO Advanced Study Institute) with the Support of the International Mathematical Union.* Academic Press, London; Thompson Book Co., Inc., Washington, DC, 1967.

[Con94] Alain Connes. *Noncommutative Geometry.* Academic Press, Inc., San Diego, CA, 1994.

[Con99] Alain Connes. Trace formula in noncommutative geometry and the zeros of the Riemann zeta function. *Selecta Mathematica*, 5(1):29, 1999.

[Dei05] Anton Deitmar. Schemes over $\mathbb{F}_1$. In *Parallel Worlds: Number Fields and Function Fields*, Progress in Mathematics, Birkhauser, pp. 87–100. 2005.

[Den18] Christopher Deninger. Dynamical systems for arithmetic schemes. *arXiv preprint arXiv:1807.06400*, 2018.

[DG71] Jean Dieudonné and Alexandre Grothendieck. *Éléments de Géométrie Algébrique*, volume 166. Springer Berlin Heidelberg New York, 1971.

[DM94] B. I. Dundas and R. McCarty. Stable K-theory and topological Hochschild homology. *Annals of Mathematics*, 140:685–701, 1994.

[DS95] W. G. Dwyer and J. Spaliński. Homotopy theories and model categories. In *Handbook of Algebraic Topology*, pp. 73–126. North-Holland, Amsterdam, 1995.

[Dur08] Nikolai Durov. New approach to Arakelov geometry. *arXiv preprint arXiv:0704.2030*, 2008.

[Fal92] Gerd Faltings. *Lectures on the Arithmetic Riemann-Roch Theorem.* Princeton University Press, 1992. Catalog Number: 127.

[Goo90] T. Goodwillie. Calculus I: The first derivative. *K-theory*, 4:1–27, 1990.

[Goo92] T. Goodwillie. Calculus II: Analytic functors. *K-theory*, 5:295–332, 1992.

[Gra76] Daniel Grayson. Higher algebraic K-theory. II (after Daniel Quillen). In *Algebraic K-theory (Proc. Conf., Northwestern Univ., Evanston, Ill., 1976)*, Lecture Notes in Math., Vol. 551, pp. 217–240. Springer, Berlin-New York, 1976.

[Gra77] Daniel R. Grayson. The K-theory of endomorphisms. *J. Algebra*, 48(2):439–446, 1977.

[Har77] Robin Hartshorne. *Algebraic Geometry*. Graduate Texts in Mathematics, No. 52. Springer-Verlag, New York-Heidelberg, 1977.

[Har89] Shai Haran. Index theory, potential theory, and the Riemann hypothesis. In *L-Functions and Arithmetic (Durham, 1989)*, volume 153 of *London Math. Soc. Lecture Note Ser.*, pp. 257–270. Cambridge University Press, Cambridge, 1989.

[Har90] Shai Haran. Riesz potentials and explicit sums in arithmetic. *Invent. Math.*, 101(3):697–703, 1990.

[Har01] Shai Haran. *The Mysteries of the Real Prime*, volume 25 of *London Mathematical Society Monographs. New Series.* The Clarendon Press, Oxford University Press, New York, 2001.

[Har07] Shai Haran. Non-additive geometry. *Compositio Mathematica*, 143:618–688, 2007.

[Har08] Shai M. J. Haran. *Arithmetical Investigations: Representation Theory, Orthogonal Polynomials, and Quantum Interpolations*, volume 1941 of *Lecture Notes in Mathematics*. Springer, 2008.

[Har09] Shai Haran. Non-additive prolegomena (to any future arithmetic that will be able to present itself as a geometry). *arXiv preprint arXiv:0911.3522*, 2009.

[Har10] Shai Haran. Invitation to nonadditive arithmetical geometry. *Casimir Force, Casimir Operators and the Riemann Hypothesis*, pp. 249–265, 2010.

[Har17a] Shai Haran. Geometry over F_1. *arXiv preprint arXiv:1709.05831*, 2017.

[Har17b] Shai Haran. *New Foundations for Geometry — Two Non-Additive Languages for Arithmetical Geometry*, volume 246 of *Memoirs of the American Mathematical Society*. 2017.

[Har20] Shai Haran. Algebra over generalized rings. *arXiv preprint arXiv:2006.15613*, 2020.

[Har22] Shai Haran. Homotopy and arithmetic. *arXiv preprint arXiv:2204.03107*, 2022.

[Hes15] Lars Hesselholt. The big de Rham-Witt complex. *Acta Math.*, 214:135–207, 2015.

[Hes17] Lars Hesselholt. Topological Hochschild homology and the Hasse-Weil zeta function. *arXiv:1602.01980v2*, 2017.

[Hov99] Mark Hovey. *Model categories*, volume 63 of *Mathematical Surveys and Monographs*. American Mathematical Society, Providence, RI, 1999.

[Hov01] Mark Hovey. Spectra and symmetric spectra in general model categories. *J. Pure Appl. Algebra*, 165(1):63–127, 2001.

[HSS00] Mark Hovey, Brooke Shipley, and Jeff Smith. Symmetric spectra. *J. Amer. Math. Soc.*, 13(1):149–208, 2000.

[Ill71] Luc Illusie. *Complexe cotangent et déformations. I.* Lecture Notes in Mathematics, Vol. 239. Springer-Verlag, Berlin-New York, 1971.

[JL01] Jay Jorgenson and Serge Lang. The ubiquitous heat kernel. In *Mathematics Unlimited—2001 and Beyond*, pp. 655–683. Springer, Berlin, 2001.

[Joy02] André Joyal. Quasi-categories and Kan complexes. *J. Pure Appl. Algebra*, 175:207–222, 2002.

[KOW02] Nobushige Kurokawa, Hiroyuki Ochiai, and Masato Wakayama. *Absolute Derivations and Zeta Functions.* Department of Math., Fac. of Science, Tokyo Inst. of Technology, 2002.

[KS18] Robert A. Kucharczyk and Peter Scholze. Topological realisations of absolute Galois groups. In James W. Cogdell, Günter Harder, Stephen Kudla, and Freydoon Shahidi, editors, *Cohomology of Arithmetic Groups*, pp. 201–288, Cham, 2018. Springer International Publishing.

[Lan78] Serge Lang. *Cyclotomic Fields.* Graduate Texts in Mathematics, Vol. 59. Springer-Verlag, New York-Heidelberg, 1978.

[Lan98] Saunders Mac Lane. *Categories for the Working Mathematician*, volume 5 of *Graduate Texts in Mathematics.* Springer-Verlag, New York, second edition, 1998.

[Lor16] Oliver Lorscheid. *A Blueprint View on* $\mathbb{F}_1$ *Geometry.* European Mathematical Society Publishing House, 2016.

[Lur04] Jacob Lurie. *Derived Algebraic Geometry.* PhD thesis, Massachusetts Institute of Technology, 2004.

[Lur09] Jacob Lurie. *Higher Topos Theory*, volume 170 of *Annals of Mathematics Studies.* Princeton University Press, Princeton, NJ, 2009.

[Lur17] Jacob Lurie. Higher algebra. 2017.

[Mac95] I. G. Macdonald. *Symmetric Functions and Hall Polynomials.* Oxford Mathematical Monographs. The Clarendon Press, Oxford University Press, New York, second edition, 1995. With contributions by A. Zelevinsky, Oxford Science Publications.

[Mad95] I.B. Madsen. *Algebraic K-Theory and Traces.* Current developments in mathematics, International Press of Boston. 1995.

[Man] Yuri Manin. Two approaches to geometries in characteristic 1: Combinatorics vs homotopy theory. Trinity College Dublin.

[Man95] Yuri Manin. Lectures on zeta functions and motives (according to Deninger and Kurokawa). *Astérisque*, 228(4):121–163, 1995.

[Onn06] Uri Onn. From p-adic to real Grassmannians via the quantum. *Adv. Math.*, 204(1):152–175, 2006.

[PL11] J. López Pena and Oliver Lorscheid. Mapping F_1-land: An overview of geometries over the field with one element. *Noncommutative Geometry, Arithmetic, and Related Topics*, Proceedings of the 21 meeting of the Japan - U.S. Mathematics Institute, pp. 241–265, 2011.

[PW92] T. Pirashvili and F. Waldhausen. Maclane homology and topological Hochschild homology. *J. Pure Appl. Alg.*, 82:81–98, 1992.

[Qui67] Daniel G. Quillen. *Homotopical Algebra.* Lecture Notes in Mathematics, No. 43. Springer-Verlag, Berlin-New York, 1967.

[Qui69] Daniel Quillen. Rational homotopy theory. *Ann. Math.*, 90(2):205–295, 1969.

[Qui70] Daniel Quillen. On the (co-) homology of commutative rings. In *Proc. Symp. Pure Math*, AMS Publications, volume 17, pp. 65–87, 1970.

[Qui72] Daniel Quillen. On the cohomology and K-theory of the general linear groups over a finite field. *Ann. Math. (2)*, 96:552–586, 1972.

[Qui73] Daniel Quillen. Higher algebraic K-theory. I. In *Algebraic K-theory, I: Higher K-theories (Proc. Conf., Battelle Memorial Inst., Seattle, Wash., 1972)*, Lecture Notes in Math., Vol. 341, pp. 85–147. Springer, Berlin-New York, 1973.

[Ram18] Srinivasa Ramanujan. On certain trigonometrical sums and their applications in the theory of numbers. *Trans. Cambridge Philos. Soc*, 22(13):259–276, 1918.

[SAB94] Christophe Soulé, Dan Abramovich, and JF Burnol. *Lectures on Arakelov Geometry.* Cambridge university press, Catalog Number: 33, 1994.

[Seg74] Graeme Segal. Categories and cohomology theories. *Topology*, 13:293–312, 1974.

[Sou08] Christophe Soulé. Les variétés sur le corps a un élément. *arXiv preprint math/0304444*, 2008.

[Tho85] R. Thomason. Algebraic K-theory and étale cohomology. *Ann. Sci. École Norm. Sup.*, 18:437–552, 1985.

[Tit57] Jacques Tits. Sur les analogues algébriques des groupes semi-simples complexes. *Centre Belge de Recherches Mathématiques Établissement Centalek*, pp. 261–289, 1957.

[TT90] R. Thomason and T. Trobaugh. Higher algebraic K-theory of schemes and of derived categories. 88:247–435, 1990.

[TV05] Bertrand Toën and Michel Vaquié. Under spec $\mathbb{Z}$. *J. K-Theory*, 3(3):437–500, 2005.

[Wal85] F. Waldhausen. *Algebraic K-theory of Spaces*, volume 1126 of *Lecture Notes in Mathematics*. Springer, 1985.

[Wei39] André Weil. Sur l'analogie entre les corps de nombres algébriques et les corps de fonctions algébriques. *Revue Scient*, 77:104–106, 1939.

[Wei66] André Weil. Fonction zêta et distributions. *Séminaire Bourbaki*, 9(312):523–531, 1966.

[Wei95] André Weil. *Basic Number Theory*. Classics in Mathematics. Springer-Verlag, Berlin, 1995. Reprint of the second (1973) edition.

[YJ15] Donald Yau and Mark W. Johnson. *A foundation for PROPs, algebras, and modules*, volume 203 of *Mathematical surveys and monographs*. American Mathematical Society, Providence, RI, 2015.

Index